EXPRESS REVIEW GUIDES

Algebra II

EXPRESS REVIEW GUIDES

Algebra II

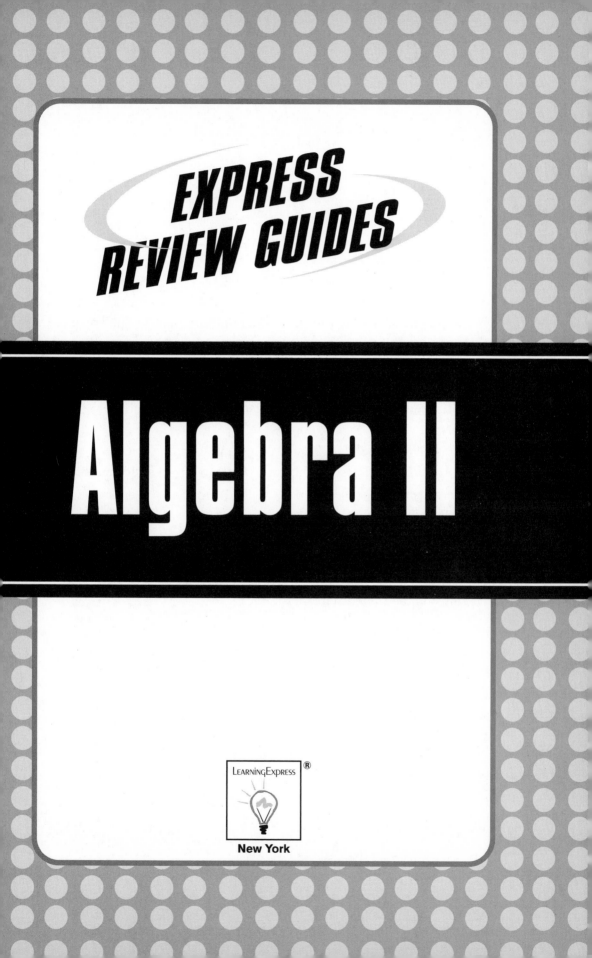

LEARNINGEXPRESS®

New York

Library of Congress Cataloging-in-Publication Data:
Express review guides. Algebra II.—1st ed.
   p. cm.
  ISBN: 978-1-57685-595-9
  1. Algebra—Outlines, syllabi, etc. I. LearningExpress (Organization) II. Title: Algebra II.
QA159.2.E97 2007
512—dc22

                                                                2007008906

Printed in the United States of America

9   8   7   6   5   4   3   2   1

First Edition

For more information or to place an order, contact LearningExpress at:
    55 Broadway
    8th Floor
    New York, NY 10006

Or visit us at:
    www.learnatest.com

Contents

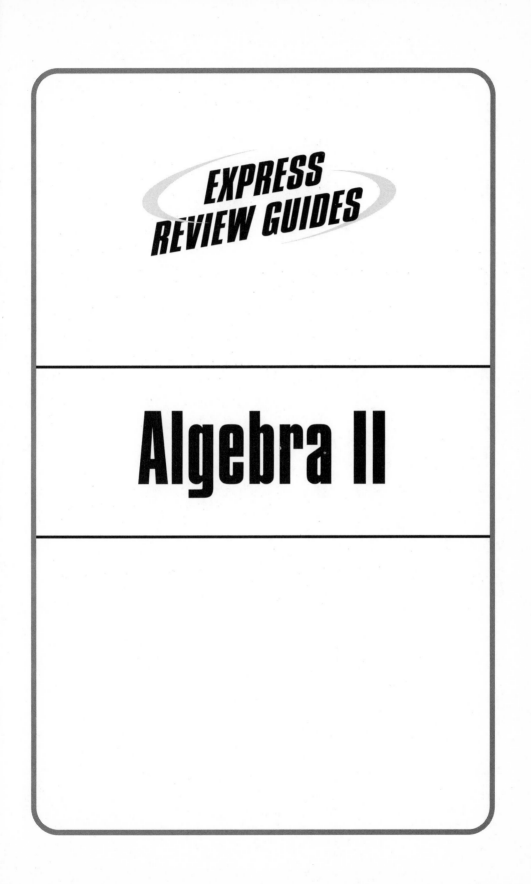

EXPRESS REVIEW GUIDES

Algebra II

Math? Why, oh Why?

Here's a common scenario that teachers run into. In every math classroom, without a doubt, a student will ask, "But why do we need to learn math?" Or, better yet, "Will we *ever* use this stuff in real life?"

Well, the answer is yes—I swear! If you cook a big meal for your family, you rely on math to follow recipes. Decorating your bedroom? You'll need math for figuring out what size bookshelf will fit in your room. Oh, and that motorcycle you want to buy? You're going to need math to figure out how much you can pay every month toward the total payment.

People have been using math for thousands of years, across various countries and continents. Whether you're sailing a boat off the coast of the Dominican Republic or building an apartment in Moscow, you're using math to get things done. You might be asking, how can math be so universal? Well, if you think about it, human beings didn't invent math concepts—we discovered them.

And then, there's algebra. Learning algebra is a little like learning another language. It's not exactly like English or Spanish or Japanese. Algebra is a simple language, used to create mathematical models of real-world situations and to help you with problems you aren't able to solve using plain old arithmetic. Rather than using words, algebra relies on numbers, symbols, and variables to make statements about things. And because algebra uses the same symbols as arithmetic for adding, subtracting, multiplying, and dividing, you're already familiar with the basic vocabulary!

People from different areas of the world and different backgrounds can "speak" algebra. If you are well versed in this language, you can use it to make important decisions and perform daily tasks with ease. Algebra can help you to shop intelligently on a budget, understand population growth, or even bet on the horse with the best chance of winning the race.

OKAY, BUT WHY *EXPRESS REVIEW GUIDES*?

If you're having trouble in math, this book can help you get on the right track. Even if you feel pretty confident about your math skills, you can use this book to help you review what you've already learned. Many study guides tell you how to improve your math—this book doesn't just *tell* you how to solve math problems, it *shows* you how. You'll find page after page of strategies that work, and you are never left stranded, wondering how to get the right answer to a problem. We'll show you all the steps to take so that you can successfully solve every single problem, and see the strategies at work.

Sometimes, math books assume you can follow explanations of difficult concepts even when they move quickly or leave out steps. That's not the case with this book. In each lesson, you'll find step-by-step strategies for tackling the different kinds of math problems. Then, you're given a chance to apply what you've learned by tackling practice problems. Answers to the practice problems are provided at the end of each section, so you can check your progress as you go along. This book is like your own personal math tutor!

THE GUTS OF THIS GUIDE

Okay, you've obviously cracked open the cover of this book if you're reading these words. But let's take a quick look at what is lurking in the other chapters. This book includes:

➡ a 50-question benchmark pretest to help you assess your knowledge of the concepts and skills in this guide
➡ brief, focused lessons covering essential algebra topics, skills, and applications
➡ specific tips and strategies to use as you study

➥ a 50-question posttest followed by complete answer explanations to help you assess your progress

As you work through this book, you'll notice that the chapters are sprinkled with all kinds of helpful tips and icons. Look for these icons and the tips they provide. They include:

➥ *Fuel for Thought*: critical information and definitions that can help you learn more about a particular topic

➥ *Practice Lap*: quick practice exercises and activities to let you test your knowledge

➥ *Inside Track*: tips for reducing your study and practice time—without sacrificing accuracy

➥ *Caution!*: pitfalls to be on the lookout for

➥ *Pace Yourself*: try these extra activities for added practice

Ready, Set, Go!

To best use this guide, you should start by taking the pretest. You'll test your math skills and see where you might need to focus your study.

Your performance on the pretest will tell you several important things. First, it will tell you how much you need to study. For example, if you got eight out of ten questions right (not counting lucky guesses!), you might need to brush up only on certain areas of knowledge. But if you got only five out of ten questions right, you will need a thorough review. Second, it can tell you what you know well (that is, which subjects you *don't* need to study). Third, you will determine which subjects you need to study in-depth and which subjects you simply need to refresh your knowledge.

After the pretest, begin the lessons, study the example problems, and try the practice problems. Check your answers as you go along, so if you miss a question, you can study a little more before moving on to the next lesson.

REMEMBER. . .

THE PRETEST IS only practice. If you did not do as well as you antic-
ipated, do not be alarmed and certainly do not despair. The purpose
of the quiz is to help you focus your efforts so that you can *improve*.
It is important to analyze your results carefully. Look beyond your score,
and consider *why* you answered some questions incorrectly. Some
questions to ask yourself when you review your wrong answers:

➡ Did you get the question wrong because the material was
   totally unfamiliar?
➡ Was the material familiar, but you were unable to come up with
   the right answer? In this case, when you read the right answer, it
   will often make perfect sense. You might even think, "I knew that!"
➡ Did you answer incorrectly because you read the question
   carelessly?

Next, look at the questions you got right and review how you came
up with the right answer. Not all right answers are created equal.

➡ Did you simply know the right answer?
➡ Did you make an educated guess? An educated guess might
   indicate that you have some familiarity with the subject, but you
   probably need at least a quick review.
➡ Did you make a lucky guess? A lucky guess means that you don't
   know the material and you will need to learn it.

After you've completed all the lessons in the book, try the posttest to see
how much you've learned. You'll also be able to see any areas where you may
need a little more practice. You can go back to the section that covers that
skill for some more review and practice.

THE RIGHT TOOLS FOR THE JOB

BE SURE THAT you have all the supplies you need on hand before you sit down to study. To help make studying more pleasant, select supplies that you enjoy using. Here is a list of supplies that you will probably need:

➥ A notebook or legal pad
➥ Graph paper
➥ Pencils
➥ Pencil sharpener
➥ Highlighter
➥ Index or other note cards
➥ Paper clips or sticky note pads for marking pages
➥ Calendar or personal digital assistant (which you will use to keep track of your study plan)
➥ Calculator

As you probably realize, no book can possibly cover all of the skills and concepts you may be faced with. However, this book is not just about building an algebra base, but also about building those essential skills that can help you solve unfamiliar and challenging questions. The algebra topics and skills in this book have been carefully selected to represent a cross section of basic skills that can be applied in a more complex setting, as needed.

Pretest

By taking this pretest, you will get an idea of how much you already know and how much you need to learn about advanced algebra.

This pretest consists of 50 questions and should take about one hour to complete. You should not use a calculator when taking this pretest; however, you may use scratch paper for your calculations. You will also need graph paper.

When you have completed this pretest, compare your answers to the answers at the end of this chapter. If your answers differ from the correct answers, use the explanations given to retrace your calculations. Each answer also tells you which chapter of this book teaches you the math skills needed for that question.

1. Given the parabola $y = 3x^2 - 3x + 1$, what is its vertex?

2. What is the line of symmetry of the parabola with equation $y = 2x^2 + 16x + 1$?

3. What is the line of symmetry of the parabola with equation $y = 3x^2$?

4. $a^3 \sqrt{(a^3)} =$

5. $\dfrac{4\sqrt{g}}{\sqrt{4g}} =$

6. $\dfrac{\sqrt{9(pr)}}{(pr)^{\frac{-3}{2}}} =$

7. $\dfrac{((\frac{x}{y})^2(\frac{y}{x})^{-2})}{xy} =$

8. What is the value of $((xy)^y)^x$ if $x = 2$ and $y = -x$?

9. If $g\sqrt{108} = \dfrac{\sqrt{3}}{g}$, what is the value of g?

10. If $\left(\dfrac{\sqrt{n}}{n^{\frac{-1}{2}}}\right)m = 5$, what is the value of m in terms of n?

11. Graph the inequality: $x > 5$

12. Graph the inequality: $y < -4$

13. Graph the following inequality: $-4x + 2y \geq 6$

14. Graph the following linear equation: $y = x + 4$

15. Graph the following linear equation: $y = 2x + 3$

16. Use scratch paper and graph paper to solve the following system of inequalities:
$y < x + 2$
$y < -x + 4$

17. Use scratch paper and graph paper to solve the following system of inequalities:
$x + y > 5$
$-2x + y > 3$

18. Find the two possible values of x that make this equation true:
$(x + 4)(x - 2) = 0$

19. Solve for x: $2x^2 - 33 = -1$

20. If $6\sqrt{d} - 10 = 32$, what is the value of d?

21. Each term in the following sequence is $\frac{3}{2}$ more than the previous term. What is the eighth term of the sequence?

$6, 7\frac{1}{2}, 9, 10\frac{1}{2}, \ldots$

22. Each term in the following sequence is seven less than the previous term. What is the value of $x - y$?

$12, 5, x, y, -16, \ldots$

23. Each term in the following sequence is $\frac{2}{3}$ times the previous term. What is the seventh term of the sequence?

$18, 12, 8, \frac{16}{3}, \ldots$

24. Each term in the following sequence is five times the previous term. What is the 20th term of the sequence?

$\frac{1}{125}, \frac{1}{25}, \frac{1}{5}, 1, \ldots$

Use the following matrix to answer questions 25 and 26.

$$A = \begin{bmatrix} 2 & 1 & 4 \\ 3 & -5 & 8 \end{bmatrix}$$

$$B = \begin{bmatrix} -3 & 2 & -1 \\ 4 & -1 & 5 \end{bmatrix}$$

25. Find $A + B$.

26. Find $A - B$.

27. Each term in the following sequence is nine less than the previous term. What is the ninth term of the sequence?

$101, 92, 83, 74, \ldots$

28. Each term in the following sequence is six more than the previous term. What is the value of $x + z$?

$x, y, z, 7, 13, \ldots$

29. Each term in the following sequence is $\frac{1}{3}$ more than the previous term. What is the value of $a + b + c + d$?

2, a, b, 3, c, d, 4, . . .

30. Each term in the following sequence is -2 times the previous term. What is the seventh term of the sequence?

3, -6, 12, -24, . . .

31. Given the following equations, what is the value of x?

$2x + y = 6$

$\frac{y}{2} + 4x = 12$

32. Given the following equations, what is the value of b?

$5a + 3b = -2$

$5a - 3b = -38$

33. Given the following equations, what is one possible value of y?

$xy = 32$

$2x - y = 0$

34. Given the following equations, what is the value of x?

$3(x + 4) - 2y = 5$

$2y - 4x = 8$

35. Given the following equations, what is the value of b?

$-7a + \frac{b}{4} = 25$

$b + a = 13$

36. Given the following equations, what is the value of y?

$3x + 7y = 19$

$\frac{4y}{x} = 1$

37. Given the following equations, what is the value of n?

$2(m + n) + m = 9$

$3m - 3n = 24$

38. Given the following equations, what is the value of b?

$9a - 2(b + 4) = 30$

$4.5a - 3b = 3$

39. Given the following equations, what is one possible value of p?

$4pq - 6 = 10$

$4p - 2q = -14$

40. Given the following equations, what is the value of a?

$7(2a + 3b) = 56$

$b + 2a = -4$

41. Given the following equations, what is the value of y?

$\frac{1}{2}x + 6y = 7$

$-4x - 15y = 10$

42. Simplify the polynomial: $9a + 12a^2 - 5a$.

43. Simplify the polynomial: $\frac{(3a)(4a)}{6(6a^2)}$.

44. Simplify the polynomial: $\frac{(5a + 7b)b}{b + 2b}$.

45. Simplify the polynomial: $(2x^2)(4y^2) + 6x^2y^2$.

46. If $a - 12 = 12$, $a = $?

47. If $6p \geq 10$, find the value of p.

48. If $x + 10 = 5$, $x = $?

49. If $\frac{k}{8} = 8$, $k = $?

50. If $-3n < 12$, find the value of n.

ANSWERS

1. First, compare the given equation to the $y = ax^2 + bx + c$ formula:

   $y = ax^2 + bx + c$

   $y = 3x^2 - 3x + 1$

   The a and the c are clear, but to clearly see what b equals, convert the subtraction to add the opposite:

   $y = 3x^2 + (-3)x + 1$

   Thus, $a = 3$, $b = -3$, and $c = 1$. The x-coordinate of the turning point, or vertex, of the parabola is given by:

   $x = \frac{-b}{2a}$

   Substitute in the values from the equation:

   $x = \frac{-b}{2a} = \frac{-(-3)}{2(3)} = \frac{3}{6} = \frac{1}{2} = 0.5$

   When $x = 0.5$, y will be:

   $y = (3)(0.25) - (3)(0.5) + 1$

   $= 0.75 - 1.5 + 1$

   $= -0.75 + 1$

   $= 0.25$

   Thus, the coordinates of the vertex are $(0.5, 0.25)$. For more help with this concept, see Chapters 7 and 8.

2. First, compare the given equation to the $y = ax^2 + bx + c$ formula:

   $y = ax^2 + bx + c$

   $y = 2x^2 + 16x + 1$

   Thus, $a = 2$, $b = 16$, and $c = 1$. The line of symmetry is given by:

   $x = \frac{-b}{2a}$

   Substitute in the values from the equation:

   $x = \frac{-b}{2a} = \frac{-16}{(2)(2)} = \frac{-16}{4} = -4$

   Thus, the line of symmetry is $x = -4$. For more help with this concept, see Chapters 7 and 8.

3. First, compare the given equation to the $y = ax^2 + bx + c$ formula:

   $y = ax^2 + bx + c$

   $y = 3x^2 + 0(x) + 0$

   Thus, $a = 3$, $b = 0$, and $c = 0$. The line of symmetry of the parabola is given by:

   $x = \frac{-b}{2a} = \frac{0}{(2)(3)} = 0$

Thus, the line of symmetry is $x = 0$. For more help with this concept, see Chapters 7 and 8.

4. Factor $\sqrt{(a^3)}$ into two radicals; a^2 is a perfect square, so factor $\sqrt{a^3}$ into $\sqrt{a}\sqrt{a^2} = a\sqrt{a}$. Multiply the coefficient of the given expression by $a\sqrt{a}$: $(a^3)(a\sqrt{a}) = a^4\sqrt{a}$. For more help with this concept, see Chapter 7.

5. Factor $\sqrt{4g}$ into two radicals; 4 is a perfect square, so factor $\sqrt{4g}$ into $\sqrt{4}\sqrt{g} = 2\sqrt{g}\sqrt{g}$. Simplify the fraction by dividing the numerator by the denominator. Cancel the $\sqrt{g}$ terms from the numerator and denominator. That leaves $\frac{4}{2} = 2$. For more help with this concept, see Chapter 7.

6. First, find the square root of $9pr$. $\sqrt{9pr} = \sqrt{9}\sqrt{pr} = 3\sqrt{pr}$. The denominator $(pr)^{\frac{-3}{2}}$ has a negative exponent, so it can be rewritten in the numerator with a positive exponent. $\sqrt{pr}$ can be written as $(pr)^{\frac{1}{2}}$, because a value raised to the exponent $\frac{1}{2}$ is another way of representing the square root of the value. The expression is now $3(pr)^{\frac{1}{2}}(pr)^{\frac{3}{2}}$. To multiply the pr terms, add the exponents. $\frac{1}{2} + \frac{3}{2} = \frac{4}{2} = 2$, so $3(pr)^{\frac{1}{2}}(pr)^{\frac{3}{2}} = 3(pr)^2 = 3p^2r^2$. For more help with this concept, see Chapter 7.

7. First, square $\frac{x}{y}$: $(\frac{x}{y})^2 = \frac{x^2}{y^2}$. Next, look at the $(\frac{y}{x})^{-2}$ term. A fraction with a negative exponent can be rewritten as the reciprocal of the fraction with a positive exponent. $(\frac{y}{x})^{-2} = (\frac{x}{y})^2 = \frac{x^2}{y^2}$. Multiply the fractions in the numerator by adding the exponents of the fractions: $(\frac{x^2}{y^2})(\frac{x^2}{y^2}) = (\frac{x^4}{y^4})$. Finally, divide this fraction by xy: $\frac{(\frac{x^4}{y^4})}{xy} = (\frac{x^4}{y^4})(\frac{1}{xy}) = \frac{x^4}{xy^5} = \frac{x^3}{y^5}$. For more help with this concept, see Chapter 7.

8. If $y = -x$, then $y = -2$. Substitute 2 for x and -2 for y: $(((2)(-2))^{-2})^2 = ((-4)^{-2})^2 = (\frac{1}{16})^2 = \frac{1}{256}$. For more help with this concept, see Chapter 7.

9. First, cross multiply: $g(g\sqrt{108}) = \sqrt{3}, g^2\sqrt{108} = \sqrt{3}$. Divide both sides of the equation by $\sqrt{108}$: $g^2\sqrt{108} = \sqrt{3}, g^2 = \frac{\sqrt{3}}{\sqrt{108}}, g^2 = \frac{1}{\sqrt{36}}, g^2 = \frac{1}{6}$.

Take the square root of both sides of the equation to find the value of g:

$g^2 = \frac{1}{6}, g = \sqrt{\frac{1}{6}} = \frac{\sqrt{1}}{\sqrt{6}} = \frac{1}{\sqrt{6}}$. Simplify the fraction by multiplying it by $\frac{\sqrt{6}}{\sqrt{6}}$:

$(\frac{1}{\sqrt{6}})(\frac{\sqrt{6}}{\sqrt{6}}) = \frac{\sqrt{6}}{6}$. For more help with this concept, see Chapter 7.

10. $\sqrt{n} = n^{\frac{1}{2}}$. The n term in the denominator has a negative exponent. It can be placed in the numerator with a positive exponent, because $\frac{1}{n^{-\frac{1}{2}}} = n^{\frac{1}{2}}$. The numerator of the fraction is now $\left(n^{\frac{1}{2}}\right)\left(n^{\frac{1}{2}}\right)$ and the denominator of the fraction is 1. To multiply terms with like bases, keep the base and add the exponents: $\left(n^{\frac{1}{2}}\right)\left(n^{\frac{1}{2}}\right) = n$. Therefore, $nm = 5$, and $m = \frac{5}{n}$. For more help with this concept, see Chapter 7.

11. For $x > 5$, draw a dashed line at $x = 5$. The area to the right satisfies the condition, so shade it:

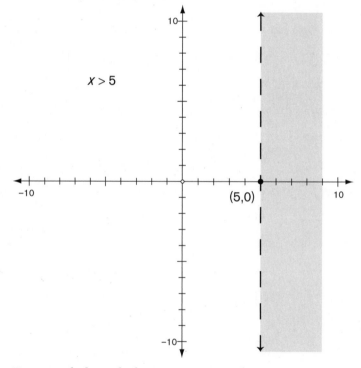

For more help with this concept, see Chapter 5.

12. For $y < -4$, draw a dashed line at $y = -4$. The area under the line satisfies the condition, so shade it:

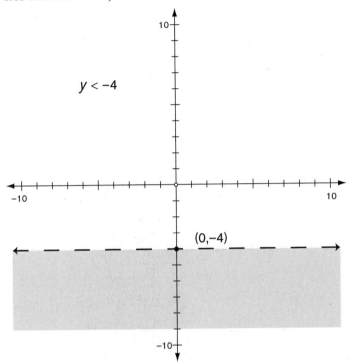

For more help with this concept, see Chapter 5.

13. For $-4x + 2y \geq 6$, rearrange so that the y is by itself on the left: $-4x + 2y \geq 6$ becomes $2y \geq 4x + 6$, which equals $y \geq 2x + 3$. For $y \geq 2x + 3$, the slope is 2 and the y-intercept is 3. Start at the y-intercept $(0,3)$ and go up 2 and over 1 (right) to plot points. This line will be solid because the symbol is $\geq$. The area above this line satisfies the condition.

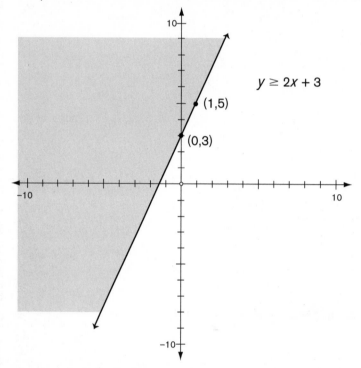

For more help with this concept, see Chapters 4 and 5.

14. For $y = x + 4$, the slope is 1 and the y-intercept is 4. Start at the y-intercept $(0,4)$ and go up 1 and over 1 (right) to plot points.

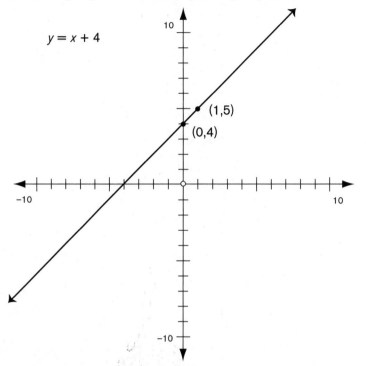

For more help with this concept, see Chapter 4.

15. For $y = 2x + 3$, the slope is 2 and the y-intercept is 3. Start at the y-intercept $(0,3)$ and go up 2 and over 1 (right) to plot points.

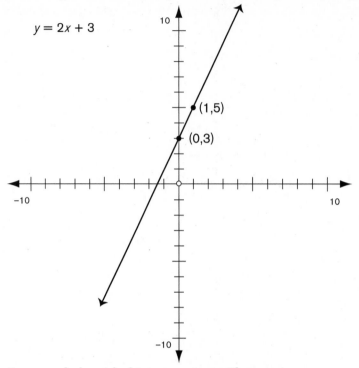

For more help with this concept, see Chapter 4.

16. For $y < x + 2$, graph the line. The slope is 1 and the y-intercept is 2. Start at the y-intercept (0,2) and go up 1 and over 1 (right) to plot points. This line will be dashed because the symbol is <. The area under this line satisfies the condition. For $y < -x + 4$, graph the line. The slope is -1 and the y-intercept is 4. Start at the y-intercept (0,4) and go down 1 and over 1 (right) to plot points. This line will also be dashed because the symbol is <. The area under this line satisfies the condition. Shade the area common to both equations:

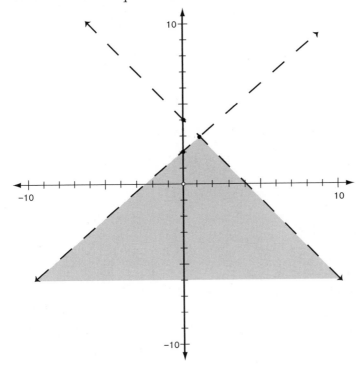

For more help with this concept, see Chapter 5.

17. Rearrange the first equation so that the y is by itself on the left: $x + y > 5$ becomes $y > -1x + 5$. For $y > -1x + 5$, the slope is -1 and the y-intercept is 5. Start at the y-intercept (0,5) and go down 1 and over 1 (right) to plot points. This line will be dashed because the symbol is >. The area above this line satisfies the condition. Rearrange the second equation so that the y is by itself on the left: $-2x + y > 3$ becomes $y > 2x + 3$. For $y > 2x + 3$, the slope is 2 and the y-intercept is 3. Start at the y-intercept (0,3) and go up 2 and over 1 (right) to plot points. This line will also be dashed because the symbol is >. The area above this line satisfies the condition. Shade the region common to both graphs:

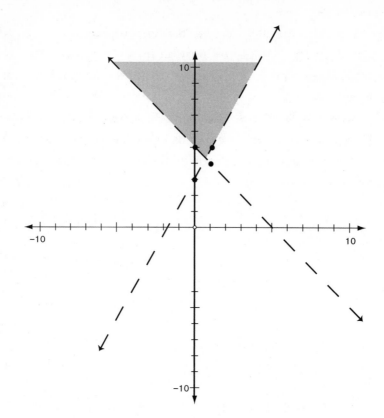

For more help with this concept, see Chapter 5.

18. Using the zero product rule, you know that either $x + 4 = 0$ or that $x - 2 = 0$.

So solve both of these possible equations:

$x + 4 = 0$ $\qquad\qquad$ $x - 2 = 0$

$x + 4 - 4 = 0 - 4$ $\qquad$ $x - 2 + 2 = 0 + 2$

$x = -4$ $\qquad\qquad\quad$ $x = 2$

So, you know that both $x = -4$ and $x = 2$ will make $(x + 4)(x - 2) = 0$. The zero product rule is useful when solving quadratic equations because you can rewrite a quadratic equation as equal to zero and take advantage of the fact that one of the factors of the quadratic equation is thus equal to 0. For more help with this concept, see Chapter 8.

19. Add 1 to both sides of the equation. $2x^2 - 33 + 1 = -1 + 1$

Simplify both sides of the equation. $2x^2 - 32 = 0$

Take out the common factor. $2(x^2 - 16) = 0$

Factor the difference of two squares. $2(x - 4)(x + 4) = 0$

Disregard the 2 and set the other factors equal to zero: $x - 4 = 0$ and $x + 4 = 0$

Solve the first equation. $x - 4 = 0$

Add 4 to both sides of the equation. $x - 4 + 4 = 0 + 4$

Simplify both sides of the equation. $x = 4$

Solve the second equation. $x + 4 = 0$

Subtract 4 from both sides of the equation. $x + 4 - 4 = 0 - 4$

Simplify both sides of the equation. $x = -4$

The solutions are 4 and -4.

For more help with this concept, see Chapters 3 and 5.

20. To solve for d, isolate the variable:

$6\sqrt{d} - 10 = 32$

$6\sqrt{d} - 10 + 10 = 32 + 10$

$6\sqrt{d} = 42$

$\frac{6\sqrt{d}}{6} = \frac{42}{6}$

$\sqrt{d} = 7$

$\sqrt{d^2} = 7^2$

$d = 49$

For more help with this concept, see Chapter 7.

21. The fourth term in the sequence is $10\frac{1}{2}$. You are looking for the eighth term, which is four terms after the fourth term. Because each term is $\frac{3}{2}$ more than the previous term, the eighth term will be $4(\frac{3}{2}) = 6$ more than $10\frac{1}{2}$; $10\frac{1}{2} + 6 = 16\frac{1}{2}$. Because the number of terms is reasonable, you can check your answer by repeatedly adding $\frac{3}{2}$: $10\frac{1}{2} + \frac{3}{2} = 12$, $12 + \frac{3}{2} = 13\frac{1}{2}$, $13\frac{1}{2} + \frac{3}{2} = 15$, $15 + \frac{3}{2} = 16\frac{1}{2}$. For more help with this concept, see Chapter 9.

22. The term that precedes x is 5. Therefore, the value of x is $5 - 7 = -2$, and the value of y is $-2 - 7 = -9$. Therefore, $x - y = -2 - (-9) = -2 + 9 = 7$. For more help with this concept, see Chapter 9.

23. The fourth term in the sequence is $\frac{16}{3}$. You are looking for the seventh term, which is three terms after the fourth term. You must multiply by $\frac{2}{3}$ three times, so the seventh term will be $(\frac{2}{3})^3 = \frac{8}{27}$ times $\frac{16}{3}$: $(\frac{8}{27})(\frac{16}{3}) = \frac{128}{81}$. Alternatively, every term in the sequence is 18 times $\frac{2}{3}$ raised to a power. The first term, 18, is $18 \times (\frac{2}{3})^0$. The second term, 12, is $18 \times (\frac{2}{3})^1$.

The value of the exponent is one less than the position of the term in the sequence. The seventh term of the sequence is equal to $18 \times$ $(\frac{2}{3})^6 = 18 \times (\frac{64}{729}) = 2 \times (\frac{64}{81}) = \frac{128}{81}$. For more help with this concept, see Chapter 9.

24. Every term in the sequence is 5 raised to a power. The first term, $\frac{1}{125}$, is 5^{-3}. The second term, $\frac{1}{25}$, is 5^{-2}. The value of the exponent is four less than the position of the term in the sequence. The 20th term of the sequence is equal to $5^{20-4} = 5^{16}$. For more help with this concept, see Chapter 9.

25. $A + B = \begin{bmatrix} 2 & 1 & 4 \\ 3 & -5 & 8 \end{bmatrix} + \begin{bmatrix} -3 & 2 & -1 \\ 4 & -1 & 5 \end{bmatrix}$

$= \begin{bmatrix} 2+(-3) & 1+2 & 4+(-1) \\ 3+4 & -5+(-1) & 8+5 \end{bmatrix}$

$= \begin{bmatrix} -1 & 3 & 3 \\ 7 & -6 & 13 \end{bmatrix}$

For more help with this concept, see Chapter 6.

26. $A - B = \begin{bmatrix} 2 & 1 & 4 \\ 3 & -5 & 8 \end{bmatrix} - \begin{bmatrix} -3 & 2 & -1 \\ 4 & -1 & 5 \end{bmatrix}$

$= \begin{bmatrix} 2-(-3) & 1-2 & 4-(-1) \\ 3-4 & -5-(-1) & 8-5 \end{bmatrix}$

$= \begin{bmatrix} 5 & -1 & 5 \\ -1 & -4 & 3 \end{bmatrix}$

For more help with this concept, see Chapter 6.

27. The fourth term in the sequence is 74. You are looking for the ninth term, which is five terms after the fourth term. Because each term is nine less than the previous term, the ninth term will be 5(9) = 45 less than 74: 74 − 45 = 29. Because the number of terms is reasonable, you can check your answer by repeatedly subtracting 9: 74 − 9 = 65, 65 −

9 = 56, 56 − 9 = 47, 47 − 9 = 38, 38 − 9 = 29. For more help with this concept, see Chapter 9.

28. The term that follows z is 7. Because each term is 6 more than the previous term, z must be 6 less than 7. Therefore, $z = 7 − 6 = 1$. In the same way, y is 6 less than z and x is 6 less than y: $y = 1 − 6 = −5$ and $x = −5 − 6 = −11$. The sum of $x + z$ is equal to $−11 + 1 = −10$. For more help with this concept, see Chapter 9.

29. The first term in the sequence is 2. The next term in the sequence, a, is $\frac{1}{3}$ more than 2: $2\frac{1}{3}$; b is $\frac{1}{3}$ more than a, $2\frac{2}{3}$; c is $\frac{1}{3}$ more than 3: $3\frac{1}{3}$; d is $\frac{1}{3}$ more than c, $3\frac{2}{3}$. Add the values of a, b, c, and d: $2\frac{1}{3} + 2\frac{2}{3} + 3\frac{1}{3} + 3\frac{2}{3} = 12$. For more help with this concept, see Chapter 9.

30. The fourth term in the sequence is −24. You are looking for the seventh term, which is three terms after the fourth term. You must multiply by −2 three times, so the seventh term will be $(−2)^3 = −8$ times −24: $(−24)(−8) = 192$. Because the number of terms is reasonable, you can check your answer by repeatedly multiplying by −2: $(−24)(−2) = 48$, $(48)(−2) = −96$, $(−96)(−2) = 192$. For more help with this concept, see Chapter 9.

31. Solve the first equation for y in terms of x: $2x + y = 6$, $y = 6 − 2x$. Substitute this expression for y in the second equation and solve for x:

$\frac{6 − 2x}{2} + 4x = 12$

$3 − x + 4x = 12$

$3x + 3 = 12$

$3x = 9$

$x = 3$

For more help with this concept, see Chapter 5.

32. Add the two equations together. The b terms will drop out, and you can solve for a:

$5a + 3b = −2$
$+ 5a − 3b = −38$
$\overline{ 10a = −40}$

$a = −4$

Substitute −4 for a in the first equation and solve for b:

$5(−4) + 3b = −2$

$−20 + 3b = −2$

$3b = 18$

$b = 6$

For more help with this concept, see Chapter 5.

33. Solve the first equation for x in terms of y: $xy = 32$, $x = \frac{32}{y}$. Substitute this expression for x in the second equation and solve for y:

$2x - y = 0$

$2(\frac{32}{y}) - y = 0$

$\frac{64}{y} - y = 0$

$\frac{64}{y} = y$

$y^2 = 64$

$y = -8, y = 8$

For more help with this concept, see Chapter 5.

34. In the first equation, multiply the $(x + 4)$ term by 3: $3(x + 4) = 3x + 12$. Then, subtract 12 from both sides of the equation, and the first equation becomes $3x - 2y = -7$. Add the two equations together, reordering them so the variables line up. The y terms will drop out, and you can solve for a:

$3x - 2y = -7$

$\underline{-4x + 2y = 8}$

$-x = 1$

$x = -1$

For more help with this concept, see Chapter 5.

35. Solve the second equation for a in terms of b: $b + a = 13$, $a = 13 - b$. Substitute this expression for a in the first equation and solve for b:

$-7a + \frac{b}{4} = 25$

$-7(13 - b) + \frac{b}{4} = 25$

$7b - 91 + \frac{b}{4} = 25$

$\frac{29b}{4} = 116$

$29b = 464$

$b = 16$

For more help with this concept, see Chapter 5.

36. Solve the second equation for x in terms of y: $\frac{4y}{x} = 1$, $x = 4y$. Substitute this expression for x in the first equation and solve for y:

$3x + 7y = 19$

$3(4y) + 7y = 19$

$12y + 7y = 19$

$19y = 19$

$y = 1$

For more help with this concept, see Chapter 5.

37. In the first equation, multiply the $(m + n)$ term by 2 and add m: $2(m + n) + m = 2m + 2n + m = 3m + 2n$. Subtract the second equation from the first equation. The m terms will drop out, and you can solve for n:

$\begin{array}{r} 3m + 2n = 9 \\ \underline{-3m - 3n = 24} \\ 5n = -15 \end{array}$

$n = -3$

For more help with this concept, see Chapter 5.

38. In the first equation, multiply the $(b + 4)$ term by -2: $-2(b + 4) = -2b - 8$. Add 8 to both sides of the equation, and the first equation becomes $9a - 2b = 38$. Multiply the second equation by 2 and subtract it from the first equation. The a terms will drop out, and you can solve for b:

$2(4.5a - 3b = 3) = 9a - 6b = 6$

$\begin{array}{r} 9a - 2b = 38 \\ \underline{-9a - 6b = 6} \\ 4b = 32 \end{array}$

$b = 8$

For more help with this concept, see Chapter 5.

39. Solve the second equation for q in terms of p: $4p - 2q = -14$, $-2q = -4p - 14$, $q = 2p + 7$. Substitute this expression for q in the first equation and solve for p:

$4pq - 6 = 10$

$4p(2p + 7) - 6 = 10$

$8p^2 + 28p - 6 = 10$

$8p^2 + 28p - 16 = 0$

$2p^2 + 7p - 4 = 0$

$(2p - 1)(p + 4) = 0$

$2p - 1 = 0, 2p = 1, p = \frac{1}{2}$

$p + 4 = 0, p = -4$

For more help with this concept, see Chapter 5.

40. Solve the second equation for b in terms of a: $b + 2a = -4$, $b = -2a - 4$. Substitute this expression for b in the first equation and solve for a:

    $7(2a + 3(-2a - 4)) = 56$

    $7(2a + -6a - 12) = 56$

    $7(-4a - 12) = 56$

    $-28a - 84 = 56$

    $-28a = 140$

    $a = -5$

    For more help with this concept, see Chapter 5.

41. Multiply the first equation by 8 and add it to the second equation. The x terms will drop out, and you can solve for y:

    $8(\frac{1}{2}x + \ 6y = 7) = 4x + 48y = 56$

    $\qquad 4x + 48y = 56$

    $\underline{+ -4x - 15y = 10}$

    $33y = 66$

    $y = 2$

    For more help with this concept, see Chapter 5.

42. Subtract the like terms by subtracting the coefficients of the terms: $9a - 5a = 4a$. Because $4a$ and $12a^2$ are not like terms, they cannot be combined any further; $9a + 12a^2 - 5a = 12a^2 + 4a$. For more help with this concept, see Chapters 3 and 7.

43. Multiply the coefficients of the terms in the numerator, and add the exponents of the bases: $(3a)(4a) = 12a^2$. Do the same with the terms in the denominator: $6(6a^2) = 36a^2$. Finally, divide the numerator by the denominator. Divide the coefficients of the terms and subtract the exponents of the bases: $\frac{(12a^2)}{(36a^2)} = \frac{1}{3}$. For more help with this concept, see Chapters 3 and 7.

44. The terms $5a$ and $7b$ have unlike bases; they cannot be combined any further. Add the terms in the denominator: $b + 2b = 3b$. Divide the b term in the numerator by the $3b$ in the denominator: $\frac{b}{3b} = \frac{1}{3}$; $(5a + 7b)(\frac{1}{3}) = \frac{5a + 7b}{3}$. For more help with this concept, see Chapter 3.

45. Multiply $2x^2$ and $4y^2$ by multiplying the coefficients of the terms: $(2x^2)(4y^2) = 8x^2y^2$. Because $8x^2y^2$ and $6x^2y^2$ have like bases, they can be added. Add the coefficients: $8x^2y^2 + 6x^2y^2 = 14x^2y^2$. For more help with this concept, see Chapters 3 and 7.

46. To solve the equation, add 12 to both sides of the equation: $a - 12 = 12$, $a - 12 + 12 = 12 + 12$, $a = 24$. For more help with this concept, see Chapters 3 and 7.

47. To solve the inequality, divide both sides of the inequality by 6: $6p \geq 10$, $\frac{6p}{6} \geq \frac{10}{6}$, $p \geq \frac{5}{3}$. For more help with this concept, see Chapters 3 and 7.

48. To solve the equation, subtract 10 from both sides of the equation: $x + 10 = 5$, $x + 10 - 10 = 5 - 10$, $x = -5$. For more help with this concept, see Chapters 3 and 7.

49. To solve the equation, multiply both sides of the equation by 8: $\frac{k}{8} = 8$, $(8)\frac{k}{8} = (8)(8)$, $k = 64$. For more help with this concept, see Chapters 3 and 7.

50. To solve the inequality, divide both sides of the inequality by -3: $-3n < 12$, $\frac{-3n}{-3} < \frac{12}{-3}$. Remember, when multiplying or dividing both sides of an inequality by a negative number, you must reverse the inequality symbol: $n > -4$. For more help with this concept, see Chapters 3 and 7.

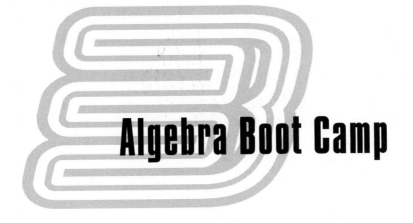

Algebra Boot Camp

Before we get into the thick of algebra II material, let's do a little algebra boot camp. This chapter reviews key skills and concepts of algebra that you will need to grasp higher-level algebra.

NUMBERS

Let's get reacquainted with the players in the game.

First, you have **whole numbers**. Whole numbers are also called "counting numbers" and include 0, 1, 2, 3, 4, 5, 6, . . . You see the pattern here.

Next on the roster are **integers**. Integers are all positive and negative whole numbers including zero: $-3, -2, -1, 0, 1, 2, 3, \ldots$

Rational numbers, good old rational numbers. Rational numbers are all numbers that can be written as fractions ($\frac{2}{3}$), terminating decimals (0.75), and repeating decimals (0.666. . .).

And rounding out the team are **irrational numbers**. Irrational numbers cannot be expressed as terminating or repeating decimals. Some examples: π or $\sqrt{2}$.

ORDER OF OPERATIONS

Most people remember the order of operations by using a mnemonic device such as **PEMDAS**, or *Please Excuse My Dear Aunt Sally*. These stand for the order in which operations are done:

Parentheses
Exponents
Multiplication
Division
Addition
Subtraction

Multiplication and division are done in the order that they appear from left to right. Addition and subtraction work the same way—left to right.

Parentheses also include any grouping symbol such as brackets [ ] and braces {}.

Examples

$$-5 + 2 \times 8 =$$
$$-5 + 16 =$$
$$11$$

$$9 + (6 + 2 \times 4) - 3^2 =$$
$$9 + (6 + 8) - 3^2 =$$
$$9 + 14 - 9 =$$
$$23 - 9 =$$
$$14$$

ABSOLUTE VALUE

The **absolute value** is the distance of a number from zero and is expressed by placing vertical bars on either side of the number. For example, $|-5|$ is 5 because -5 is 5 spaces from zero. Most people simply remember that the absolute value of a number is its positive form.

Examples

$|-39| = 39$
$|92| = 92$
$|-11| = 11$
$|987| = 987$

NUMBERS AND LETTERS MEET

Variables are letters that are used to represent numbers. Once you realize that these variables are just numbers in disguise, you'll understand that they must obey all the rules of mathematics, just like the numbers that aren't disguised. This can help you figure out what number the variable at hand stands for.

FUEL FOR THOUGHT

WHEN A NUMBER is placed next to a variable, indicating multiplication, the number is said to be the **coefficient** of the variable. For example,

$8c$ 8 is the coefficient to the variable c.

$6ab$ 6 is the coefficient to both variables a and b.

FUEL FOR THOUGHT

If two or more terms have exactly the same variable(s), they are said to be **like terms**.

$$7x + 3x = 10x$$

The process of grouping like terms together performing mathematical operations is called combining like terms. It is important to combine like terms carefully, making sure that the variables are exactly the same.

ENTER EXPRESSIONS AND EQUATIONS

An **expression** is like a series of words without a verb. Take, for example, $3x + 5$ or $a - 3$.

An equation is a statement that includes the "verb," in this case, an equal sign. To solve an algebraic **equation** with one variable, find the value of the unknown variable.

Rules for Working with Equations

1. The equal sign separates an equation into two sides.
2. Whenever an operation is performed on one side, the same operation must be performed on the other side.
3. To solve an equation, first move all of the variables to one side and all of the numbers to the other. Then simplify until only one variable (with a coefficient of 1) remains on one side and one number remains on the other side.

Example

$7x - 11 = 29 - 3x$	Move the variables to one side.
$7x - 11 + 3x = 29 - 3x + 3x$	Perform the same operation on both sides.
$10x - 11 = 29$	Now move the numbers to the other side.

$$10x - 11 + 11 = 29 + 11 \qquad \text{Perform the same operation on both sides.}$$

$$10x = 40 \qquad \text{Divide both sides by the coefficient.}$$
$$\frac{10x}{10} = \frac{40}{10} \qquad \text{Simplify.}$$
$$x = 4$$

PRACTICE LAP

DIRECTIONS: Use scratch paper to solve the following problem. You can check your answer at the end of this chapter.

1. If $13x - 28 = 22 - 12x$, what is the value of x?

CROSS PRODUCTS

You can solve an equation that sets one fraction equal to another by finding **cross products** of the fractions. Finding cross products allows you to remove the denominators from each side of the equation. Multiply each side by a fraction equal to 1 that has the denominator from the opposite side.

Example

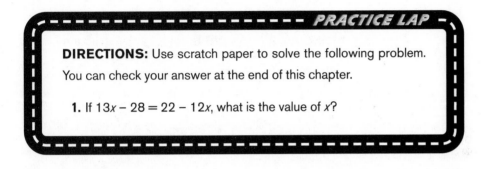

$$\frac{a}{b} = \frac{c}{d}$$

First multiply one side by $\frac{d}{d}$ and the other by $\frac{b}{b}$. The fractions $\frac{d}{d}$ and $\frac{b}{b}$ both equal 1, so they don't change the value of either side of the equation.

$$\frac{a}{b} \times \frac{d}{d} = \frac{c}{d} \times \frac{b}{b}$$
$$\frac{ad}{bd} = \frac{bc}{bd}$$

The denominators are now the same. Now multiply both sides by the denominator and simplify.

$$bd \times \frac{ad}{bd} = bd \times \frac{bc}{bd}$$
$$ad = bc$$

This example demonstrates how finding cross products works. In the future, you can skip all the middle steps and just assume that $\frac{a}{b} = \frac{c}{d}$ is the same as $ad = bc$.

Examples

$\frac{x}{6} = \frac{12}{36}$ Find cross products.

$36x = 6 \times 12$

$36x = 72$

$x = 2$

$\frac{x}{4} = x + \frac{12}{16}$ Find cross products.

$4x = 16x + 12$

$-12x = 12$

$x = -1$

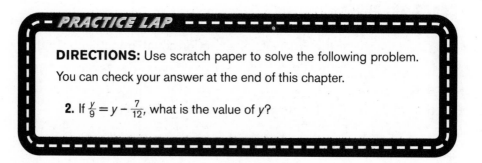

PRACTICE LAP

DIRECTIONS: Use scratch paper to solve the following problem. You can check your answer at the end of this chapter.

2. If $\frac{y}{9} = y - \frac{7}{12}$, what is the value of y?

CHECKING EQUATIONS

After you solve an equation, you can check your answer by substituting your value for the variable into the original equation.

Example

We found that the solution for $7x - 11 = 29 - 3x$ is $x = 4$. To check that the solution is correct, substitute 4 for x in the equation:

$7x - 11 = 29 - 3x$

$7(4) - 11 = 29 - 3(4)$

$28 - 11 = 29 - 12$

$17 = 17$

This equation is true, so $x = 4$ is the correct solution!

INSIDE TRACK

SPECIAL TIPS FOR CHECKING EQUATIONS ON EXAMS

IF TIME PERMITS, check all equations. For questions that ask you to find the solution to an equation, you can simply substitute each answer choice into the equation and determine which value makes the equation correct.

Be careful to answer the question that is being asked. Sometimes, questions require that you solve for a variable and then perform an operation. For example, a question may ask the value of $x - 2$. You might find that $x = 2$ and look for an answer choice of 2. But because the question asks for the value of $x - 2$, the answer is not 2, but $2 - 2$. Thus, the answer is 0.

EQUATIONS WITH MORE THAN ONE VARIABLE

Some equations have more than one variable. To find the solution of these equations, solve for one variable in terms of the other(s). Follow the same method as when solving single-variable equations, but isolate only one variable.

Example

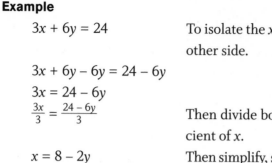

$3x + 6y = 24$ To isolate the x variable, move $6y$ to the other side.

$3x + 6y - 6y = 24 - 6y$

$3x = 24 - 6y$

$\frac{3x}{3} = \frac{24 - 6y}{3}$ Then divide both sides by 3, the coefficient of x.

$x = 8 - 2y$ Then simplify, solving for x in terms of y.

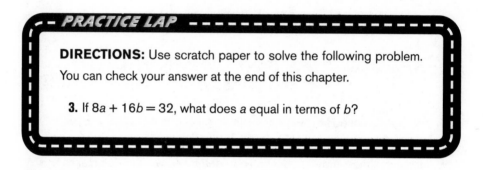

- - PRACTICE LAP -

DIRECTIONS: Use scratch paper to solve the following problem. You can check your answer at the end of this chapter.

3. If $8a + 16b = 32$, what does a equal in terms of b?

INEQUALITIES

An inequality is two numbers or expressions that are connected with an inequality symbol. Inequalities contain the greater than, less than, greater than or equal to, less than or equal, or not equal to symbols. When you solve the inequalities for x, you can figure out a range of numbers that your unknown is allowed to be.

This symbol . . .	Means . . .
$>$	"Greater than"
$\geq$	"Greater than or equal to"
$<$	"Less than"
$\leq$	"Less than or equal to"
$\neq$	"Not equal to"

Here are some examples of inequalities:

$2 < 5$ (2 is less than 5)

$9 > 3$ (9 is greater than 3)

$4 \leq 4$ (4 is less than or equal to 4)

$2x + 5 \neq 11$ (2x added to 5 is not equal to 11)

Now, there is one rule that you need to remember when dealing with inequalities: When you multiply or divide by a negative number, you need to reverse the sign.

INSIDE TRACK

WHENEVER YOU MULTIPLY or divide an inequality by a negative number, you need to reverse the inequality symbol.

Example

$-5x + 3 > 28$ can also be expressed as what inequality?

The goal here is to isolate your x. First, subtract 3 from both sides.

$-5x + 3 - 3 > 28 - 3$

$-5x \qquad\quad > 25$

When you multiply or divide by a negative number, you need to reverse the sign. When you divide by -5, you get:

$\frac{-5x}{-5} > \frac{25}{-5}$

So, $x < -5$.

INSIDE TRACK

HERE'S A TRICK to avoid having to worry about flipping the sign: Just move your terms in a manner such that you will end up with a *positive* coefficient on your variable.

Let's look at $-5x + 3 > 28$ again. Our x coefficient is negative, so you will add $+5x$ to both sides.

$-5x + 3 + 5x > 28 + 5x$

$ 3 > 28 + 5x$

Next, subtract 28 from both sides:

$3 - 28 > 28 + 5x - 28$

$3 - 28 > 5x$

$ -25 > 5x$

You divide both sides by 5 to yield:

$-5 > x$

SOLVING INEQUALITIES

You can solve inequalities with variables just like you can solve equations with variables. Use what you already know about solving equations to solve inequalities. Like equations, you can add, subtract, multiply, or divide both sides of an inequality with the same number. In other words, what you do to one side of an inequality, you must do to the other side.

Example

$2x + 3 < 1$

Subtract 3 from both sides of the inequality.	$2x + 3 - 3 < 1 - 3$
Simplify both sides of the inequality.	$2x < -2$
Divide both sides of the inequality by 2.	$\frac{2x}{2} < \frac{-2}{2}$
Simplify both sides of the inequality.	$x < -1$

The answer for this example is the inequality $x < -1$. There is an endless number of solutions because every number less than -1 is an answer. In this problem, the number -1 is not an answer because the inequality states that your answers must be numbers less than -1.

FUEL FOR THOUGHT

THE ANSWER TO an inequality will always be an inequality. Because the answer is an inequality, you will have an infinite number of solutions.

SOLVING INEQUALITIES VERSUS SOLVING EQUATIONS

Did you notice the similarity between solving equations and solving inequalities? Well, there are some major *differences* you need to be aware of.

Notice what happens when you multiply or divide an inequality by a *negative* number.

$$
\begin{array}{ccc}
2 & < & 5 \\
-2 \cdot 2 & < & 5 \cdot -2 \\
-4 & < & -10
\end{array}
$$

However, -4 is not less than -10. So, $-4 < -10$ is a false statement. To correct it, you would have to rewrite it as $-4 > -10$.

You can solve inequalities using the same methods that you use to solve equations with these exceptions:

- When you multiply or divide an inequality by a negative number, you must reverse the inequality symbol.
- The answer to an inequality will always be an inequality.

USING VARIABLES TO
EXPRESS RELATIONSHIPS

THE MOST IMPORTANT skill for solving word problems is being able to use variables to express relationships. This list will assist you in this by giving you some common examples of English phrases and their mathematical equivalents.

➥ "Increase" means add.
  "Decrease" means subtract.

Example

A number increased by five $= x + 5$.
A number decreased by five $= x - 5$.

➥ "Less than" means subtract.
  "More than" means add.

Example

Ten less than a number $= x - 10$.
Ten more than a number $= x + 10$.

➥ "Times" or "product" means multiply.
  "Divisible" or "quotient" means divide.

Example

Three times a number $= 3x$.
Three is divisible by a number $= 3 \div x$.

➥ "Times the sum" means to multiply a number by a quantity.

Example

Five times the sum of a number and three $= 5(x + 3)$.

➥ Two variables are sometimes used together.

Example

A number y exceeds five times a number x by ten.

$y = 5x + 10$

➥ Inequality signs are used for "at least" and "at most," as well as "less than" and "more than."

Examples

The product of x and 6 is greater than 2.

$x \cdot 6 > 2$

When 14 is added to a number x, the sum is less than 21.

$x + 14 < 21$

The sum of a number x and four is at least nine.

$x + 4 \geq 9$

When seven is subtracted from a number x, the difference is at most four.

$x - 7 \leq 4$

ABSOLUTE VALUE INEQUALITIES

$|x| < a$ is equivalent to $-a < x < a$, and $|x| > a$ is equivalent to $x > a$ or $x < -a$.

Example

$|x + 3| > 7$

$x + 3 > 7 \quad \text{or} \quad x + 3 < -7$
$x > 4 \qquad\qquad x < -10$

Thus, $x > 4$ or $x < -10$.

ANSWERS

1. To solve for x:

   $13x - 28 = 22 - 12x$

   $13x - 28 + 12x = 22 - 12x + 12x$

   $25x - 28 = 22$

   $25x - 28 + 28 = 22 + 28$

   $25x = 50$

   $x = 2$

2. To solve for y:

   $\frac{y}{9} = y - \frac{7}{12}$

   $4y = 36y - 21$

   $-32y = -21$

   $y = \frac{21}{32}$

3. To solve for a in terms of b:

   $8a + 16b = 32$

   $8a + 16b - 16b = 32 - 16b$

   $8a = 32 - 16b$

   $\frac{8a}{8} = 32 - \frac{16b}{8}$

   $a = 4 - 2b$

Functions and Graphs

function is a relationship in which one value depends upon another value. For example, if you are buying candy bars at a certain price, there is a relationship between the number of candy bars you buy and the amount of money you have to pay.

Think about functions as a machine—you put something into the machine, and it spits something back out. For example, when you enter quarters into a vending machine, you get a snack. (That is, unless it gets stuck on the ledge.) This is like a function. The input to the function was quarters, and the output of the function was a snack.

Basically, a function is a set of rules for using input and to produce output, and usually, this involves numbers.

Functions are written in the form beginning with the following symbol:

$f(x) =$

For example, consider the function $f(x) = 8x - 2$. If you are asked to find $f(3)$, you simply substitute the 3 into the given function equation.

$f(x) = 8x - 2$
becomes
$f(3) = 8(3) - 2$
$f(3) = 24 - 2 = 22$

So, when $x = 3$, the value of the function is 22.

- - PRACTICE LAP

DIRECTIONS: Use scratch paper to solve the following problem. You can check your answer at the end of this chapter.

1. Using the function, $f(x) = 5x - 1$, find $f(3)$.

You could also imagine functions that take more than one number as their input, like $f(x,y) = x + y$. That means that if you give the function the numbers 7 and 4 as input, the function spits out the number 11 as output.

Function tables portray a relationship between two variables, such as an x and a y. It is your job to figure out exactly what that relationship is. Let's look at a function table:

x	y
0	—
1	4
2	5
3	—
4	7

Notice that some of the data was left out. Don't worry about that! You can still figure out what you need to do to the x in order to make it the y. You see that $x = 1$ corresponds to $y = 4$; $x = 2$ corresponds to $y = 5$; and $x = 4$ corresponds to $y = 7$. Did you spot the pattern? Our y-value is just our x-value plus 3.

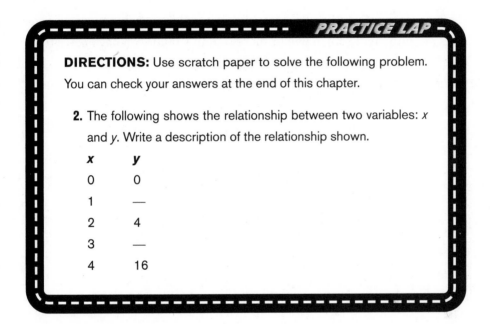

PRACTICE LAP

DIRECTIONS: Use scratch paper to solve the following problem. You can check your answers at the end of this chapter.

2. The following shows the relationship between two variables: x and y. Write a description of the relationship shown.

x	y
0	0
1	—
2	4
3	—
4	16

FUNCTIONS AND COORDINATE GRIDS

Okay, here's a quick review of the coordinate grid.

In a coordinate grid, the horizontal axis is the **x-axis**, and the vertical axis is the **y-axis**. The place where they meet is the point of origin. Using this system, you can place any point on the grid if you give it an x-value and a y-value, conventionally written as (x,y). If you have two or more points, you have a **line**.

Now think about a basic function. If you input an initial value x, you get an $f(x)$ value. If you call $f(x)$ the y-value, you can see how a typical function can spit out a huge number of points that can then be graphed. Look back at $f(x) = 8x - 2$. If $x = 2$, $y = 14$. If $x = 3$, $y = 22$. So, we already have two points for this line: $(2,14)$ and $(3,22)$. Once you know two points of a function, you can draw a line connecting them. And guess what? You have now graphed a function!

Potential functions must pass the **vertical line test** in order to be considered a function. The vertical line test is the following: Does a vertical line drawn through a graph of the potential function pass through only one point of the graph? If YES, then the vertical line passes through only one point, and the potential function is a function. If NO, then the vertical line passes through more than one point, and the potential function is *not* a function.

The **horizontal line test** can be used to determine how many different *x*-values, or inputs, return the same *f(x)*-value. Remember, a function cannot have one input return two or more outputs, but it can have more than one input return the same output. For example, the function $f(x) = x^2$ is a function, because no *x*-value can return two or more *f(x)* values, but more than one *x*-value can return the same *f(x)*-value. Both $x = 2$ and $x = -2$ make $f(x) = 4$. To find how many values make $f(x) = 4$, draw a horizontal line through the graph of the function where *f(x)*, or *y*, = 4.

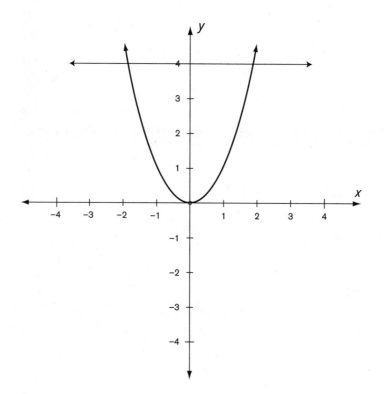

You can see that the line $y = 4$ crosses the graph of $f(x) = x^2$ in exactly two places. Therefore, the horizontal line test proves that there are two values for x that make $f(x) = 4$.

All of the x-values of a function, collectively, are called its **domain**. Sometimes there are x-values that are outside of the domain, but these are the x-values for which the function is not defined.

The function $f(x) = 3x$ has a domain of all real numbers. Any real number can be substituted for x in the equation and the value of the function will be a real number.

The function $f(x) = \frac{2}{x} - 4$ has a domain of all real numbers excluding 4. If $x = 4$, the value of the function would be $\frac{2}{0}$, which is undefined. In a function, the values that make a part of the function undefined are the values that are NOT in the domain of the function.

What is the domain of the function $f(x) = \sqrt{x}$?

The square root of a negative number is undefined, so the value of x must not be less than 0. Therefore, the domain of the function is $x \geq 0$.

All of the solutions to $f(x)$ are collectively called the **range**. Any values that $f(x)$ cannot be equal to are said to be outside of the range.

As you just saw, the function $f(x) = 3x$ has a domain of all real numbers. If any real number can be substituted for x, $3x$ can yield any real number. The range of this function is also all real numbers.

Although the domain of the function $f(x) = \frac{2}{x-4}$ is all real numbers excluding 4, the range of the function is all real numbers excluding 0, because no value for x can make $f(x) = 0$.

What is the range of the function $f(x) = \sqrt{x}$?

You already found the domain of the function to be $x \geq 0$. For all values of x greater than or equal to 0, the function will return values greater than or equal to 0.

The x-values are known as the **independent variables**. The y-values *depend* on the x-values, so the y-values are called the **dependent variables**.

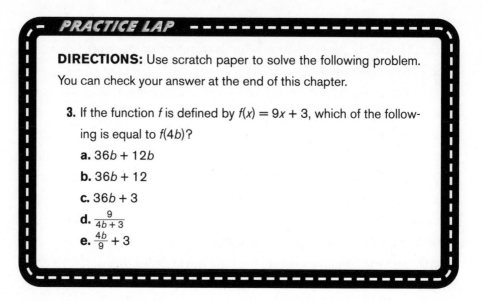

PRACTICE LAP

DIRECTIONS: Use scratch paper to solve the following problem. You can check your answer at the end of this chapter.

3. If the function *f* is defined by $f(x) = 9x + 3$, which of the following is equal to $f(4b)$?

a. $36b + 12b$

b. $36b + 12$

c. $36b + 3$

d. $\frac{9}{4b+3}$

e. $\frac{4b}{9} + 3$

NESTED FUNCTIONS

Given the definitions of two functions, you can find the result of one function (given a value) and place it directly into another function. For example, if $f(x) = 5x + 2$ and $g(x) = -2x$, what is $f(g(x))$ when $x = 3$?

Begin with the innermost function: Find $g(x)$ when $x = 3$. In other words, find $g(3)$. Then, substitute the result of that function for x in $f(x)$: $g(3) = -2(3) = -6$, $f(-6) = 5(-6) + 2 = -30 + 2 = -28$. Therefore, $f(g(x)) = -28$ when $x = 3$.

What is the value of $g(f(x))$ when $x = 3$?

Start with the innermost function—this time, it is $f(x)$: $f(3) = 5(3) + 2 = 15 + 2 = 17$. Now, substitute 17 for x in $g(x)$: $g(17) = -2(17) = -34$. When $x = 3$, $f(g(x)) = -28$ and $g(f(x)) = -34$.

NEWLY DEFINED SYMBOLS

A symbol can be used to represent one or more operations. A symbol such as # may be given a certain definition, such as "$m\#n$ is equivalent to $m^2 + n$." You may be asked to find the value of the function given the values of m and n, or you may be asked to find an expression that represents the function.

If $m\#n$ is equivalent to $m^2 + n$, what is the value of $m\#n$ when $m = 2$ and $n = -2$?

Substitute the values of m and n into the definition of the symbol. The definition of the function states that the term before the # symbol should be squared and added to the term after the # symbol. When $m = 2$ and $n = -2$, $m^2 + n = (2)^2 + (-2) = 4 - 2 = 2$.

If $m\#n$ is equivalent to $m^2 + 2n$, what is the value of $n\#m$?

The definition of the function states that the term before the # symbol should be squared and added to twice the term after the # symbol. Therefore, the value of $n\#m = n^2 + 2m$. Watch your variables carefully. The definition of the function is given for $m\#n$, but the question asks for the value of $n\#m$.

If $m\#n$ is equivalent to $m + 3n$, what is the value of $n\#(m\#n)$?

Begin with the innermost function, $m\#n$. The definition of the function states that the term before the # symbol should be added to three times the term after the # symbol. Therefore, the value of $m\#n = m + 3n$. That expression, $m + 3n$, is now the term after the # symbol: $n\#(m + 3n)$. Look again at the definition of the function. Add the term before the # symbol to three times the term after the # symbol. Add n to three times $(m + 3n)$: $n + 3(m + 3n) = n + 3m + 9n = 3m + 10n$.

USING THE SLOPE AND *y*-INTERCEPT

A linear equation always graphs into a straight line. The variable in a linear equation cannot contain an exponent greater than one. It cannot have a variable in the denominator, and the variables cannot be multiplied.

The graph of a linear equation is a line, which means it goes on forever in both directions. A graph is a picture of all the answers to the equation, so there is an infinite (endless) number of solutions. Every point on that line is a solution.

There are several methods that can be used to graph linear equations; however, you will use the slope-intercept method here.

What does slope mean to you? If you are a skier, you might think of a ski slope. The slope of a line has a similar meaning. The **slope** of a line is the steepness of a line. What is the *y*-intercept? Intercept means to cross, so the *y*-intercept is where the line crosses the *y*-axis. A positive slope will always rise from left to right. A negative slope will always fall from left to right.

To graph a linear equation, you will first change its equation into **slope-intercept form**. The slope-intercept form of a linear equation is $y = mx + b$, also known as $y =$ form. Linear equations have two variables. For example, in the equation $y = mx + b$, the two variables are x and y. The m represents a number and is the *slope* of the line and is also a constant. The b represents a number and is the **y-intercept**. For example, in the equation $y = 2x + 3$, the number 2 is the m, which is the slope. The 3 is the b, which is the y-intercept. In the equation $y = -3x + 5$, the m is -3, and the b is 5.

Okay, now you know that slope means the steepness of a line. In the equation $y = 2x + 3$, the slope of the line is 2. What does it mean when you have a slope of 2? Slope is defined as the rise of the line over the run of the line. If the slope is 2, this means $\frac{2}{1}$, so the rise is 2 and the run is 1.

If the slope of a line is $\frac{2}{3}$, the rise is 2 and the run is 3. What do rise and run mean? **Rise** is the vertical change, and **run** is the horizontal change. To graph a line passing through the origin with a slope of $\frac{2}{3}$, start at the origin. The rise is 2, so from the origin, go up 2 and to the right 3. Then draw a line from the origin to the endpoint. The line you have drawn has a slope of $\frac{2}{3}$.

Now draw a line with a slope of $-\frac{3}{4}$. Start at the origin. Go down 3 units because you have a negative slope. Then go right 4 units. Finally, draw a line from the origin to the endpoint. These two lines appear on the same graph that follows.

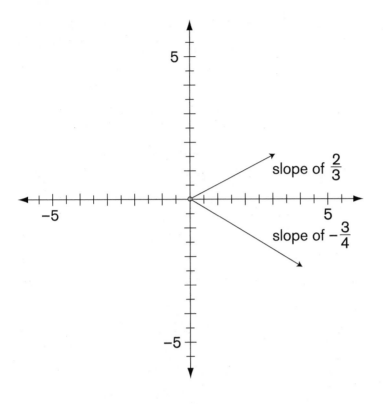

Let's try another one. To graph an equation like $y = x + 1$, you can use the slope and y-intercept. The first step is to figure out what the slope is. The slope is the number in front of x, which means in this case that it is 1. What is the y-intercept? It is also 1. To graph the equation, your starting point will be the y-intercept, which is 1. From the y-intercept, use the slope, which is also 1, or $\frac{1}{1}$. The slope tells you to go up 1 and to the right 1. A line is drawn from the y-intercept to the endpoint (1,2). You can extend this line and draw arrows on each end to show that the line extends infinitely.

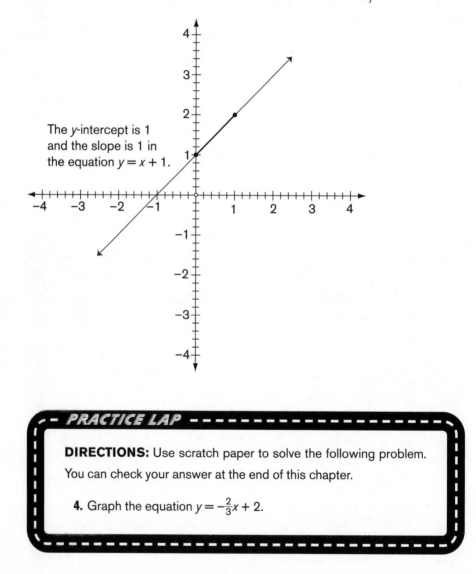

The y-intercept is 1 and the slope is 1 in the equation $y = x + 1$.

PRACTICE LAP

DIRECTIONS: Use scratch paper to solve the following problem. You can check your answer at the end of this chapter.

4. Graph the equation $y = -\frac{2}{3}x + 2$.

GETTING THE RIGHT FORM

What if the equation is not in slope-intercept form? Simple! All you need to do is change the equation to slope-intercept form. How? Slope-intercept form is y = form, so your strategy is to get the y on a side by itself.

An equation needs to be in slope-intercept form, or y = form ($y = mx + b$), before you can graph the equation with a pencil and graph paper. Also, if you use a graphing calculator to graph a linear equation, the equation needs to be in y = form before it can be entered into the calculator.

Example

$2x + y = 5$

Subtract $2x$ from both sides of the equation.	$2x - 2x + y = 5 - 2x$
Simplify.	$y = 5 - 2x$
Rearrange the equation so the x term is first.	$y = -2x + 5$

There is a mathematical rule called the **commutative property** that lets you change the order of numbers or terms when you add or multiply. You want the preceding equation in the form $y = mx + b$, so the order of the 5 and the $-2x$ needs to be changed after getting the y on a side by itself. When you move a term, be sure to take the sign of the term with it. For example, the 5 was a positive number in the original order. It remains a positive number when you move it.

Example

$2x + 3y = 9$

Subtract $2x$ from both sides of the equation.	$2x - 2x + 3y = 9 - 2x$
Simplify.	$3y = 9 - 2x$
Use the commutative property.	$3y = -2x + 9$
Divide both sides by 3.	$\frac{3y}{3} = -\frac{2x}{3} + \frac{9}{3}$
Simplify both sides of the equation.	$y = -\frac{2x}{3} + 3$

Example

$$-3x + 2y = 10$$

Add $3x$ to both sides of the equation.	$-3x + 3x + 2y = 10 + 3x$
Simplify.	$2y = 10 + 3x$
Use the commutative property.	$2y = 3x + 10$
Divide both sides of the equation by 2.	$\frac{2y}{2} = \frac{3x}{2} + \frac{10}{2}$
Simplify both sides of the equation.	$y = \frac{3}{2}x + 5$

THE DEAL WITH LINEAR INEQUALITIES

A **linear inequality** has two variables just like a linear equation. The inequality $2x + y < 1$ is a linear inequality with two variables. You can draw on what you already know to graph linear inequalities. A linear equation graphs into a line. A linear inequality has two parts: a line and a shaded area.

When you graphed linear equations, your first step was to put the equation into $y =$ form. Do the same with the linear inequality. The commutative property lets you change the order of numbers or terms when you add or multiply. When you move a term, be sure to take the sign of the term with it.

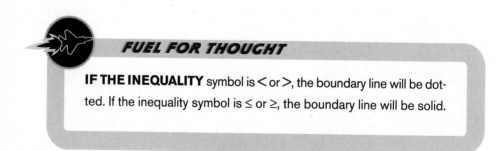

Example

$2x + y > 1$

Subtract 2x from both sides of the inequality. $2x - 2x + y > 1 - 2x$

Simplify. $y > 1 - 2x$

Use the commutative property. $y > -2x + 1$

The inequality $y > -2x + 1$ tells you that the slope is -2 and the y-intercept is 1. If the inequality has a $>$ or $<$ symbol, then the line will be dotted. If the inequality symbol is $\leq$ or $\geq$, then the line will be solid. Generally, if the inequality symbol is $>$ or $\geq$, you shade above the line. If the inequality symbol is $<$ or $\leq$, you shade below the line.

To graph $y > -2x + 1$, start with the y-intercept, which is 1. The slope is -2, which means $-\frac{2}{1}$, so from the y-intercept of 1, go down 2, because the slope is negative, and to the right 1. When you connect the starting point and the ending point, you will have the boundary line of your shaded area. You should extend this line as far as you'd like in either direction because it is end-less. The boundary line will be dotted because the inequality symbol is $>$. If the symbol had been $\geq$, then the line would be solid, not dotted.

When you graph, always check your graph to make certain the direction of the line is correct. If the slope is positive, the line should rise from left to right. If the slope is negative, the line should fall from left to right.

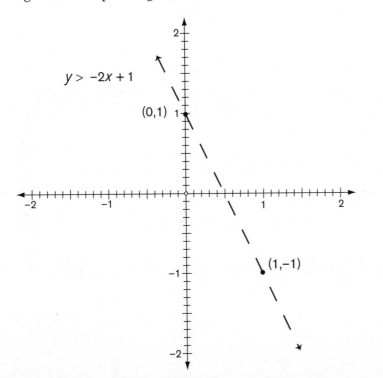

Now that you have the boundary line, will you shade above or below the line? The inequality symbol is >, so shade above the line.

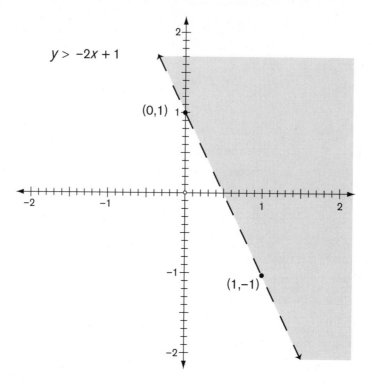

$y > -2x + 1$

(0,1)

(1,−1)

If the inequality symbol is > or ≥, you will shade above the boundary line. If the inequality symbol is < or ≤, you will shade below the boundary line.

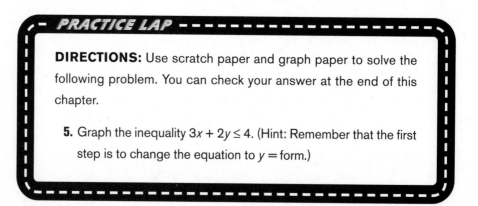

PRACTICE LAP

DIRECTIONS: Use scratch paper and graph paper to solve the following problem. You can check your answer at the end of this chapter.

5. Graph the inequality $3x + 2y \leq 4$. (Hint: Remember that the first step is to change the equation to $y =$ form.)

SPECIAL CASES OF INEQUALITIES

There are two special cases of inequalities. One has a vertical boundary line and the other has a horizontal boundary line. For example, the inequality $x > 2$ will have a vertical boundary line, and the inequality $y > 2$ will have a horizontal boundary line. The inequality $y > 2$ is the same as the inequality $y > 0x + 2$. It has a slope of 0 and a y-intercept of 2. When the slope is 0, the boundary line will always be a horizontal line.

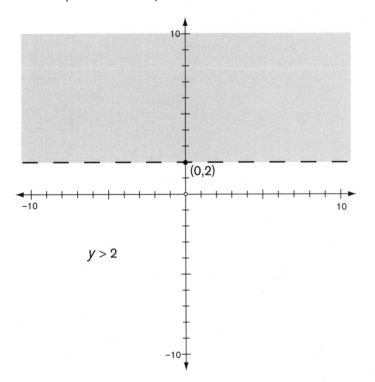

The inequality $x > 2$ cannot be written in $y =$ form because it does not have a slope or a y-intercept. It will always be a vertical line. It will be a dotted vertical line passing through the point on the x-axis where $x = 2$.

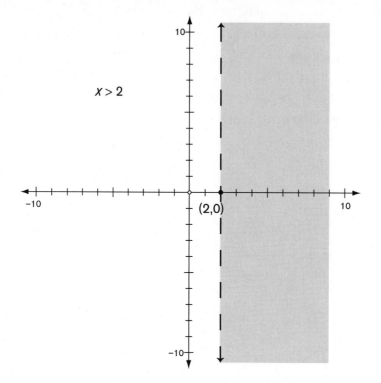

A horizontal line has a slope of 0. A vertical line has no slope.

ANSWERS

1. $f(3) = 5(3) - 1 = 15 - 1 = 14$. So, $f(3) = 14$.
2. The y-value is the x-value squared.
3. If $f(x) = 9x + 3$, then, for $f(4b)$, $4b$ simply replaces x in $9x + 3$. Therefore, $f(4b) = 9(4b) + 3 = 36b + 3$.
4. Start with the y-intercept, which is a positive 2. From there, go down 2, because the slope is negative, and to the right 3. Draw a line to connect the origin and the endpoint.

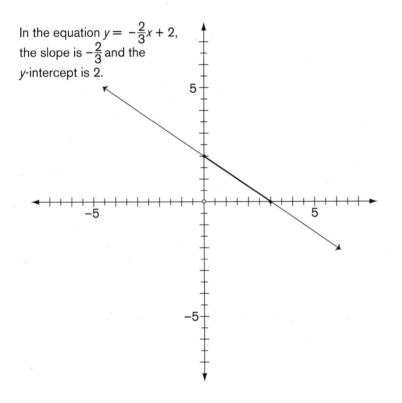

In the equation $y = -\frac{2}{3}x + 2$, the slope is $-\frac{2}{3}$ and the y-intercept is 2.

5. Subtract $3x$ from both sides of the inequality. $3x - 3x + 2y \le 4 - 3x$
   Simplify. $2y \le 4 - 3x$
   Use the commutative property. $2y \le -3x + 4$
   Divide both sides of the inequality by 2. $\frac{2y}{2} \le -\frac{3x}{2} + \frac{4}{2}$
   Simplify both sides of the inequality. $y \le -\frac{3}{2}x + 2$
   The y-intercept of the inequality is 2 and the slope is $-\frac{3}{2}$. Start with the y-intercept, which is 2. From that point, go down 3 because the slope

is negative and to the right 2. Then connect the starting point and the ending point. Will the boundary line be dotted or solid? It will be solid because the inequality symbol is ≤. Will you shade above or below the boundary line? You will shade below the boundary line because the inequality symbol is ≤.

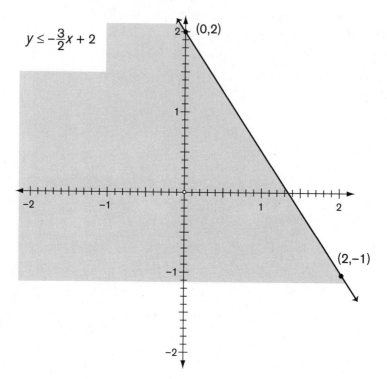

Tackling Systems of Equations and Inequalities

WHAT'S AROUND THE BEND

➡ Systems of Linear Equations
   and Inequalities
➡ Substitution
➡ Linear Combination
➡ Solving Systems of Equalities
   Graphically
➡ Graphing Systems of Inequalities

system of equations is a set of two or more equations with the same solution. For example, if we're told that both $2c + d = 11$ and $c + 2d = 13$ have the same solution, that means the variables c and d have the same value in both equations.

Two methods for solving a system of equations are **substitution** and **linear combination**.

SUBSTITUTION

Substitution involves solving for one variable in terms of another and then substituting that expression into the second equation.

Example

Here are those two equations with the same solution:

$2c + d = 11$ and $c + 2d = 13$

To solve, first choose one of the equations and rewrite it, isolating one variable in terms of the other. It does not matter which variable you choose.

$2c + d = 11$ becomes $d = 11 - 2c$

Next substitute $11 - 2c$ for d in the other equation and solve:

$c + 2d = 13$

$c + 2(11 - 2c) = 13$

$c + 22 - 4c = 13$

$22 - 3c = 13$

$22 = 13 + 3c$

$9 = 3c$

$c = 3$

Now substitute this answer into either original equation for c to find d.

$2c + d = 11$

$2(3) + d = 11$

$6 + d = 11$

$d = 5$

Thus, $c = 3$ and $d = 5$.

LINEAR COMBINATION

Linear combination, which can also be used to solve a system of equations, involves writing one equation over another and then adding or subtracting the like terms so that one letter is eliminated.

Example

$x - 7 = 3y$ and $x + 5 = 6y$

First rewrite each equation in the same form.

$x - 7 = 3y$ becomes $x - 3y = 7$

$x + 5 = 6y$ becomes $x - 6y = -5$.

Now subtract the two equations so that the x terms are eliminated, leaving only one variable:

$$x - 3y = 7$$
$$- (x - 6y = -5)$$
$$\overline{ 3y = 12}$$

$y = 4$ is the answer.

Now substitute 4 for y in one of the original equations and solve for x.

$x - 7 = 3y$

$x - 7 = 3(4)$

$x - 7 = 12$

$x - 7 + 7 = 12 + 7$

$x = 19$

So, the solution to the system of equations is $y = 4$ and $x = 19$.

SYSTEMS OF EQUATIONS WITH NO SOLUTION

It is possible for a system of equations to have no solution if there are no values for the variables that would make all the equations true. For example, the following system of equations has no solution because there are no values of x and y that would make both equations true:

$$3x + 6y = 14$$
$$3x + 6y = 9$$

In other words, the same expression cannot equal both 14 and 9.

DIRECTIONS: Use scratch paper and graph paper to solve the following system of equalities. You can check your answer at the end of this chapter.

1. $5x + 3y = 4$

$15x + dy = 21$

What value of d would give the system of equations NO solution: $-9, -3, 1,$ or 9?

GRAPHING SYSTEMS OF EQUALITIES

To graph a system of linear equations, use what you already know about graphing linear equations. To graph a system of linear equations, you will use the slope-intercept form of graphing. The first step is to transform the equations into slope-intercept form or $y = mx + b$. Then use the slope and y-intercept to graph the line. Once you have both lines graphed, determine your solutions.

Example

$x - y = 6$

$2x + y = 3$

Transform the first equation into $y = mx + b$.	$x - y = 6$
Subtract x from both sides of the equation.	$x - x - y = 6 - x$
Simplify.	$-y = 6 - x$
Use the commutative property.	$-y = -x + 6$
Multiply both sides of the equation by -1.	$-1 \cdot -y = -1(-x + 6)$
Simplify both sides.	$y = x - 6$
Transform the second equation into $y = mx + b$.	$2x + y = 3$
Subtract $2x$ from both sides of the equation.	$2x - 2x + y = 3 - 2x$
Simplify.	$y = 3 - 2x$
Use the commutative property.	$y = -2x + 3$

The slope of the first equation is 1, and the y-intercept is –6. The slope of the second equation is –2, and the y-intercept is 3. In the first equation, the line cuts the y-axis at –6. From that point, go up 1 and to the right 1. Draw a line through your beginning point and the endpoint—this line can extend as long as you want in both directions because it is endless. In the second equation, the line cuts the y-axis at 3. From that point, go down 2 and to the right 1. Draw a line through your beginning point and the endpoint, extending it as long as you want. The point of intersection of the two lines is (3,–3), so there is one solution.

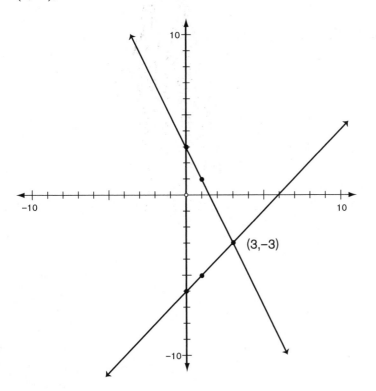

SOLVING SYSTEMS OF INEQUALITIES GRAPHICALLY

A system of inequalities is two or more inequalities with the same variables. You graph systems of linear inequalities in the same way that you graph systems of linear equations. However, remember that the graph of an inequality consists of a boundary line *and* a shaded area. Review Chapter 4 if you need help recalling the basics of graphing inequalities.

To graph a system of inequalities, transform the inequality into $y = mx + b$, and graph the boundary line. Then determine if you will shade above or below the boundary line. The solution of the system of inequalities will be the intersection of the shaded areas. Look at the following inequalities.

$$y > x$$
$$y < 3$$

The inequalities are already in $y =$ form, so you are ready to graph them. The slope of the first equation is 1, and the y-intercept is 0. Start with the y-intercept of 0 and then go up 1 and to the right 1. The boundary line will be dotted because the inequality symbol is $>$. Because the inequality is $>$, you will shade above the line.

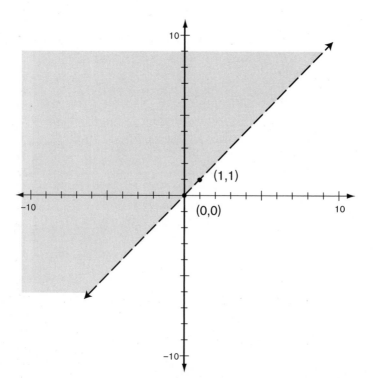

The inequality $y < 3$ will have a horizontal boundary line. The boundary line will be dotted, and you will shade below the line because the inequality symbol is $<$.

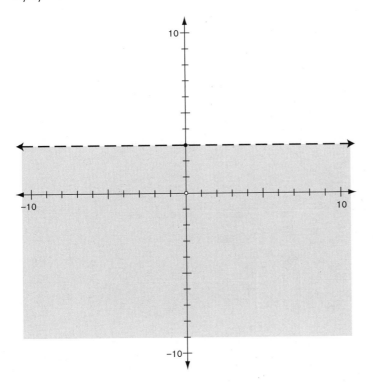

The intersection of the two shaded areas is the solution of the system.

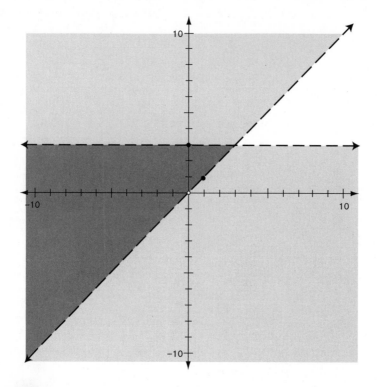

PACE YOURSELF

USE A SYSTEM of inequalities to represent how you will spend your money. Let x = the amount of money you need to spend on necessities. Let y = the amount of money you can spend on recreation. Fill in the system of inequalities with the amount of money that fits your circumstances.

$x + y \leq$ (amount of money you have to spend)
$x \geq$ (amount of money you need to spend on necessities)

When you have filled in the dollar amounts, graph the system of inequalities. You do not want any values less than zero. Why?

Example

$$-x + y \geq 4$$
$$2x + y \leq 1$$

Transform the first inequality to $y = mx + b$.	$-x + y \geq 4$
Add x to both sides of the inequality.	$-x + x + y \geq 4 + x$
Simplify.	$y \geq 4 + x$
Use the commutative property.	$y \geq x + 4$
Transform the second inequality to $y = mx + b$.	$2x + y \leq 1$
Subtract $2x$ from both sides of the inequality.	$2x - 2x + y \leq 1 - 2x$
Simplify.	$y \leq 1 - 2x$
Use the commutative property.	$y \leq -2x + 1$

The slope of the first inequality is 1 and the y-intercept is 4. To graph the inequality, start with the y-intercept, which is 4. From that point, go up 1 and to the right 1. Draw a line through the starting point and the endpoint. The boundary line will be a solid line, and you will shade above it because the inequality symbol is $\geq$.

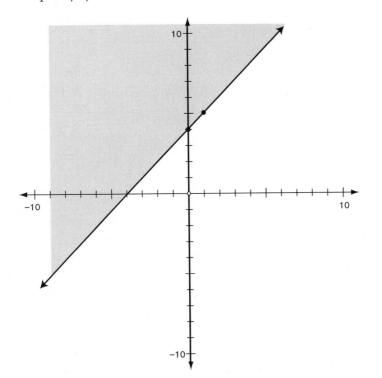

The slope of the second inequality is –2, and the *y*-intercept is 1. To graph the inequality, start with the *y*-intercept, which is 1. From that point, go down 2 and to the right 1. Draw a line through the starting point and the endpoint. The boundary line will be a solid line, and you will shade below the line because the inequality symbol is ≤.

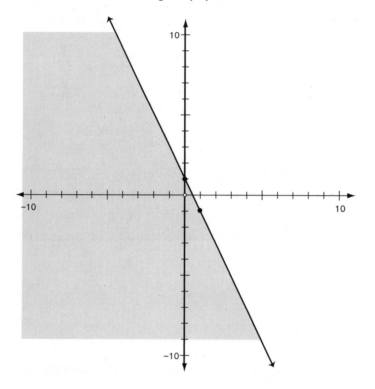

The solution of the system of inequalities is the intersection of the two shaded areas.

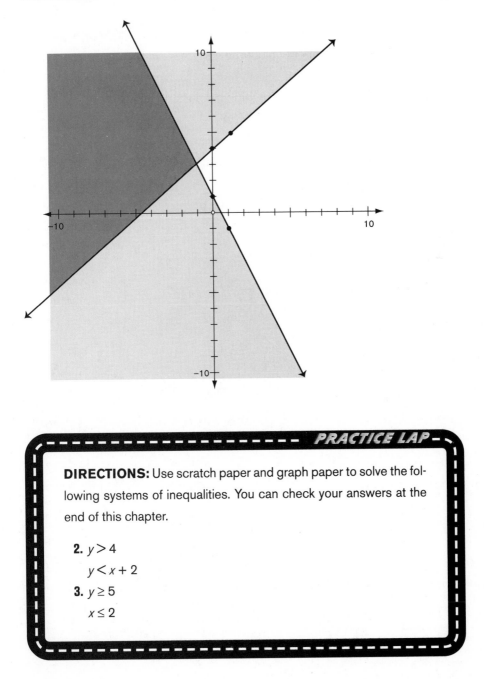

PRACTICE LAP

DIRECTIONS: Use scratch paper and graph paper to solve the following systems of inequalities. You can check your answers at the end of this chapter.

2. $y > 4$
   $y < x + 2$

3. $y \geq 5$
   $x \leq 2$

ANSWERS

1. The first step in evaluating a system of equations is to write the equations so that the coefficients of one of the variables are the same. If you multiply $5x + 3y = 4$ by 3, you get $15x + 9y = 12$. Now you can compare the two equations because the coefficients of the x variables are the same:

   $15x + 9y = 12$

   $15x + dy = 21$

   The only reason there would be no solution to this system of equations is if the system sets the same expression equal to different numbers. Therefore, you must choose the value of d that would make $15x + dy$ identical to $15x + 9y$. If $d = 9$, then:

   $15x + 9y = 12$

   $15x + 9y = 21$

   Thus, if $d = 9$, there is no solution.

2. For $y > 4$, draw a dashed line at $y = 4$. The area of the graph above this line will satisfy the condition. For $y < x + 2$, graph the line. The slope is 1 and the y-intercept is 2. Start at the y-intercept $(0,2)$ and go up 1 and over 1 (right) to plot points. This line will also be dashed because the symbol is $<$. The area under this line satisfies the condition. Shade the area where both conditions are satisfied:

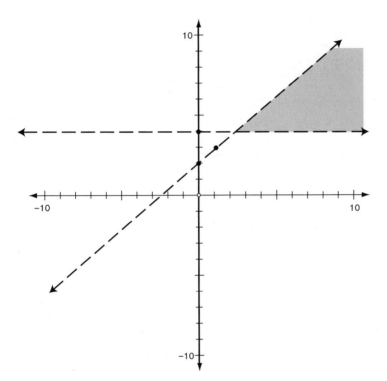

3. For $y \geq 5$, draw a solid line at $y = 4$. The area of the graph above this line will satisfy the condition. For $x \leq 2$, draw a solid line at $x = 2$. The area of the graph to the left of this line will satisfy the condition. Shade the area common to both conditions:

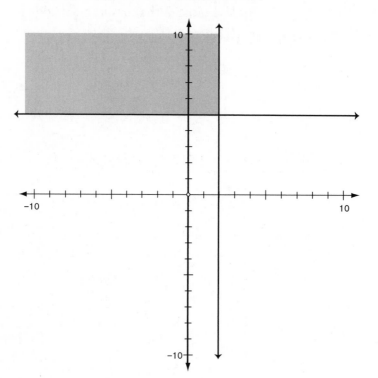

6

Demystifying Matrices

WHAT'S AROUND THE BEND

➥ Definition of a Matrix
➥ Adding and Subtracting Matrices
➥ Multiplying Matrices
➥ Special Matrices

You may think you know about the matrix; however, in algebra, the matrix is not another world where you can perform magical stunts and fight bad guys.

In algebra, matrices are rectangular arrays of numbers. Look at the example of a 2-by-2 matrix:

$$\begin{bmatrix} a_1 & a_2 \\ a_3 & a_4 \end{bmatrix}$$

The brackets indicate that it's a matrix. The horizontal rows are called the **rows** and the vertical columns are simply called **columns**. The numbers that appear are called the entries or elements of the matrix.

Matrices are not limited to two rows by two columns, though. Look at the following example.

$$\begin{bmatrix} m & n & o & p \\ q & r & s & t \end{bmatrix}$$

This matrix is a 2-by-4 matrix; it has two rows and four columns.

rows: $[m\,n\,o\,p]$ and $[q\,r\,s\,t]$

columns: $\begin{bmatrix} m, & n, & o, & p \\ q & r & s & t \end{bmatrix}$

m is the 1st row, 1st column entry.
n is the 1st row, 2nd column entry.
o is the 1st row, 3rd column entry.
p is the 1st row, 4th column entry.
q is the 2nd row, 1st column entry.
r is the 2nd row, 2nd column entry.
s is the 2nd row, 3rd column entry.
t is the 2nd row, 4th column entry.

To perform operations on matrices, there are a few basic rules to follow.

ADDITION

To add two matrices of the same kind, simply add the corresponding entries. To add matrices, both must have the same number of rows and columns. Be careful to add the same entry from both matrices.

$$\begin{bmatrix} a_1 & a_2 \\ a_3 & a_4 \end{bmatrix} + \begin{bmatrix} b_1 & b_2 \\ b_3 & b_4 \end{bmatrix} = \begin{bmatrix} a_1 + b_1 & a_2 + b_2 \\ a_3 + b_3 & a_4 + b_4 \end{bmatrix}$$

Because matrices are added by adding corresponding entries, matrix addition is commutative and associative. In short, if A, B, and C are matrices:

$A + B = B + A$

And $A + (B + C) = (A + B) + C$

SUBTRACTION

Subtraction follows the same rule as addition.

$$\begin{bmatrix} a_1 & a_2 \\ a_3 & a_4 \end{bmatrix} - \begin{bmatrix} b_1 & b_2 \\ b_3 & b_4 \end{bmatrix} = \begin{bmatrix} a_1 - b_1 & a_2 - b_2 \\ a_3 - b_3 & a_4 - b_4 \end{bmatrix}$$

MULTIPLICATION

If the columns in the first matrix are the same as the rows of the second matrix, you can multiply two matrices. You'll get a matrix that is the row of the first matrix by the column of the second matrix. Note: You cannot multiply them in reverse order.

$$\begin{bmatrix} a_1 & a_2 \\ a_3 & a_4 \end{bmatrix} \times \begin{bmatrix} b_1 & b_2 \\ b_3 & b_4 \end{bmatrix} = \begin{bmatrix} a_1 b_1 + a_2 b_3 & a_1 b_2 + a_2 b_4 \\ a_3 b_1 + a_4 b_3 & a_3 b_2 + a_4 b_4 \end{bmatrix}$$

To multiply a matrix with a real number, multiply each element with this real number.

$$k \begin{bmatrix} a_1 & a_2 \\ a_3 & a_4 \end{bmatrix} = \begin{bmatrix} ka_1 & ka_2 \\ ka_3 & ka_4 \end{bmatrix}$$

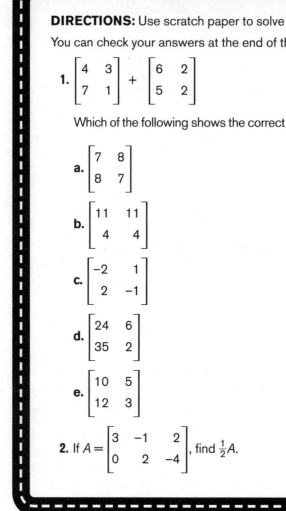

PRACTICE LAP

DIRECTIONS: Use scratch paper to solve the following problems. You can check your answers at the end of this chapter.

1. $\begin{bmatrix} 4 & 3 \\ 7 & 1 \end{bmatrix} + \begin{bmatrix} 6 & 2 \\ 5 & 2 \end{bmatrix}$

Which of the following shows the correct solution to the problem?

a. $\begin{bmatrix} 7 & 8 \\ 8 & 7 \end{bmatrix}$

b. $\begin{bmatrix} 11 & 11 \\ 4 & 4 \end{bmatrix}$

c. $\begin{bmatrix} -2 & 1 \\ 2 & -1 \end{bmatrix}$

d. $\begin{bmatrix} 24 & 6 \\ 35 & 2 \end{bmatrix}$

e. $\begin{bmatrix} 10 & 5 \\ 12 & 3 \end{bmatrix}$

2. If $A = \begin{bmatrix} 3 & -1 & 2 \\ 0 & 2 & -4 \end{bmatrix}$, find $\frac{1}{2}A$.

SPECIAL MATRICES

Some matrices consist of a single row:

$A = \begin{bmatrix} 2 & 5 & 1 & 7 & 0 \end{bmatrix}$

This **row matrix** would be referred to as a 1-by-5 matrix.

Other matrices consist of a single column. Look at the following **column matrix**. It would be called a 5-by-1 matrix.

$$B = \begin{bmatrix} 4 \\ 1 \\ 6 \\ 9 \\ 0 \end{bmatrix}$$

If the number of entries in A is the same as the elements in B, the product AB of A and B is the number obtained by pairing each entry of A with the corresponding element of B, multiplying these pairs, and adding the resulting products.

Example

Find CD if $C = \begin{bmatrix} 2 & 8 & 3 \end{bmatrix}$ and $D = \begin{bmatrix} 60 \\ 20 \\ 300 \end{bmatrix}$

$$CD = \begin{bmatrix} 2 & 8 & 3 \end{bmatrix} \begin{bmatrix} 60 \\ 20 \\ 300 \end{bmatrix} = 2(60) + 8(20) + 3(300) = 1{,}180$$

PRACTICE LAP

DIRECTIONS: Use scratch paper to solve the following problem. You can check your answer at the end of this chapter.

3. Find CD if $C = \begin{bmatrix} 4 & 2 & 3 \end{bmatrix}$ and $D = \begin{bmatrix} 25 \\ 56 \\ 110 \end{bmatrix}$

ANSWERS

1. $\begin{bmatrix} 4 & 3 \\ 7 & 1 \end{bmatrix} + \begin{bmatrix} 6 & 2 \\ 5 & 2 \end{bmatrix} = \begin{bmatrix} 4+6 & 3+2 \\ 7+5 & 1+2 \end{bmatrix} = \begin{bmatrix} 10 & 5 \\ 12 & 3 \end{bmatrix}$

2. $\frac{1}{2}A = \frac{1}{2} \begin{bmatrix} 3 & -1 & 2 \\ 0 & 2 & -4 \end{bmatrix}$

$= \begin{bmatrix} \frac{3}{2} & -\frac{1}{2} & 1 \\ 0 & 1 & -2 \end{bmatrix}$

3. $CD = \begin{bmatrix} 4 & 2 & 3 \end{bmatrix} \begin{bmatrix} 25 \\ 56 \\ 110 \end{bmatrix} = 4(25) + 2(56) + 3(110) = 100 + 112 + 330 = 542$

7 Polynomials and Radicals— No Problem!

WHAT'S AROUND THE BEND

- ➥ Monomials
- ➥ Polynomials
- ➥ Operations and Polynomials
- ➥ FOIL
- ➥ Factoring
- ➥ Common Factors
- ➥ Isolating Variables Using Fractions
- ➥ Reciprocals
- ➥ Roots
- ➥ Undefined Expressions
- ➥ Radicals
- ➥ Exponents

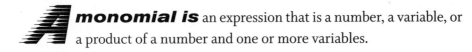 **monomial is** an expression that is a number, a variable, or a product of a number and one or more variables.

$$6 \qquad y \qquad 5xy^2 \qquad 19a^6b^4$$

A **polynomial** is a monomial or the sum or difference of two or more monomials.

$$7y^5 \qquad -6ab^4 \qquad 8x + y^3 \qquad 8x + 9y - z$$

FUEL FOR THOUGHT

HERE ARE THREE kinds of polynomials:

1. A **monomial** is a polynomial with one term, such as $5b^6$.
2. A **binomial** is a polynomial with two unlike terms, such as $2x + 4y$.
3. A **trinomial** is a polynomial with three unlike terms, such as $y^3 + 8z - 2$.

OPERATIONS WITH POLYNOMIALS

To add polynomials, simply combine like terms. Let's look at $(5y^3 - 2y + 1) + (y^3 + 7y - 4)$.

First remove the parentheses: $5y^3 - 2y + 1 + y^3 + 7y - 4$. Then arrange the terms so that like terms are grouped together: $5y^3 + y^3 - 2y + 7y + 1 - 4$. Now combine like terms: $6y^3 + 5y - 3$. Just to make sure you've got it, let's look at another example.

Example

$(2x - 5y + 8z) - (16x + 4y - 10z)$

First remove the parentheses. Be sure to distribute the subtraction correctly to all terms in the second set of parentheses:
$2x - 5y + 8z - 16x - 4y + 10z$
Then arrange the terms so that like terms are grouped together:
$2x - 16x - 5y - 4y + 8z + 10z$
Now combine like terms:
$-14x - 9y + 18z$

To multiply monomials, multiply their coefficients and multiply like variables by adding their exponents.

Example

$$(-4a^3b)(6a^2b^3) = (-4)(6)(a^3)(a^2)(b)(b^3) = -24a^5b^4$$

To divide monomials, divide their coefficients and divide like variables by subtracting their exponents.

Example

$$\frac{10x^5y^7}{15x^4y^2} = \left(\frac{10}{15}\right)\left(\frac{x^5}{x^4}\right)\left(\frac{y^7}{y^2}\right) = \frac{2xy^5}{3}$$

To multiply a polynomial by a monomial, multiply each term of the polynomial by the monomial and add the products.

Example

$$8x(12x - 3y + 9)$$

Distribute. $(8x)(12x) - (8x)(3y) + (8x)(9)$

Simplify. $96x^2 - 24xy + 72x$

To divide a polynomial by a monomial, divide each term of the polynomial by the monomial and add the quotients.

Example

$$\frac{6x - 18y + 42}{6} = \frac{6x}{6} - \frac{18y}{6} + \frac{42}{6} = x - 3y + 7$$

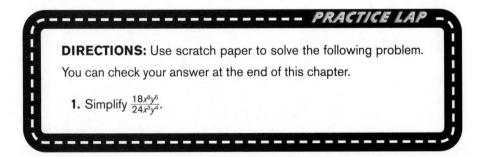

PRACTICE LAP

DIRECTIONS: Use scratch paper to solve the following problem. You can check your answer at the end of this chapter.

1. Simplify $\frac{18x^8y^5}{24x^3y^4}$.

FOIL

The FOIL method is used when multiplying binomials. FOIL represents the order used to multiply the terms: First, Outer, Inner, and Last. To multiply binomials, you multiply according to the FOIL order and then add the products.

Example

$(4x + 2)(9x + 8)$

F: $4x$ and $9x$ are the **first** pair of terms.
O: $4x$ and 8 are the **outer** pair of terms.
I: 2 and $9x$ are the **inner** pair of terms.
L: 2 and 8 are the **last** pair of terms.
Now multiply according to FOIL:
$(4x)(9x) + (4x)(8) + (2)(9x) + (2)(8) = 36x^2 + 32x + 18x + 16$
Now combine like terms:
$36x^2 + 50x + 16$

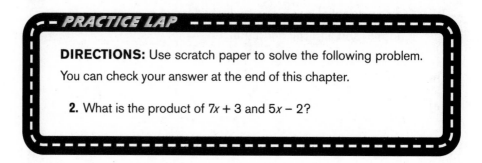

- - PRACTICE LAP - - - - - - - - - - - - - - - - - -

DIRECTIONS: Use scratch paper to solve the following problem. You can check your answer at the end of this chapter.

2. What is the product of $7x + 3$ and $5x - 2$?

FIGURING OUT FACTORING

Factoring is the reverse of multiplication. When multiplying, you find the product of factors. When factoring, you find the factors of a product.

Multiplication: $3(x + y) = 3x + 3y$
Factoring: $3x + 3y = 3(x + y)$

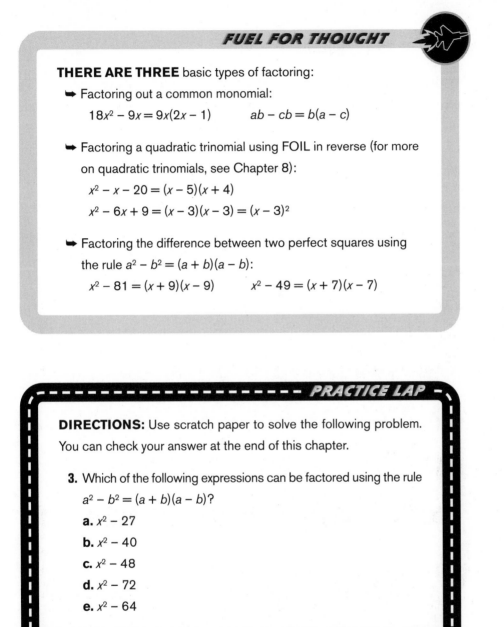

FUEL FOR THOUGHT

THERE ARE THREE basic types of factoring:

➥ Factoring out a common monomial:

$$18x^2 - 9x = 9x(2x - 1) \qquad ab - cb = b(a - c)$$

➥ Factoring a quadratic trinomial using FOIL in reverse (for more on quadratic trinomials, see Chapter 8):

$$x^2 - x - 20 = (x - 5)(x + 4)$$
$$x^2 - 6x + 9 = (x - 3)(x - 3) = (x - 3)^2$$

➥ Factoring the difference between two perfect squares using the rule $a^2 - b^2 = (a + b)(a - b)$:

$$x^2 - 81 = (x + 9)(x - 9) \qquad x^2 - 49 = (x + 7)(x - 7)$$

PRACTICE LAP

DIRECTIONS: Use scratch paper to solve the following problem. You can check your answer at the end of this chapter.

3. Which of the following expressions can be factored using the rule $a^2 - b^2 = (a + b)(a - b)$?

a. $x^2 - 27$

b. $x^2 - 40$

c. $x^2 - 48$

d. $x^2 - 72$

e. $x^2 - 64$

USING COMMON FACTORS

With some polynomials, you can determine a **common factor** for all of the terms. For example, $4x$ is a common factor of all three terms in the polynomial $16x^4 + 8x^2 + 24x$ because it can divide evenly into each of them. To factor a polynomial with terms that have common factors, you can divide the polynomial by the known factor to determine the second factor.

Example

In the binomial $64x^3 + 24x$, $8x$ is the greatest common factor of both terms. Therefore, you can divide $64x^3 + 24x$ by $8x$ to find the other factor.

$$\frac{64x^3 + 24x}{8x} = \frac{64x^3}{8x} + \frac{24x}{8x} = 8x^2 + 3$$

Thus, factoring $64x^3 + 24x$ results in $8x(8x^2 + 3)$.

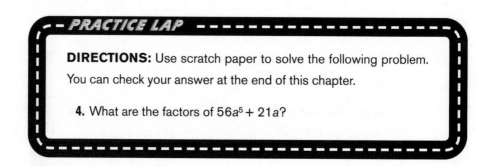

DIRECTIONS: Use scratch paper to solve the following problem. You can check your answer at the end of this chapter.

4. What are the factors of $56a^5 + 21a$?

ISOLATING VARIABLES USING FRACTIONS

It may be necessary to use factoring in order to isolate a variable in an equation.

Example

If $ax - c = bx + d$, what is x in terms of a, b, c, and d?

First isolate the x terms on the same side of the equation:

$ax - bx = c + d$

Now factor out the common x term:

$x(a - b) = c + d$

Then divide both sides by $a - b$ to isolate the variable x:

$$\frac{x(a-b)}{a-b} = \frac{c+d}{a-b}$$

Simplify:

$$x = \frac{c+d}{a-b}$$

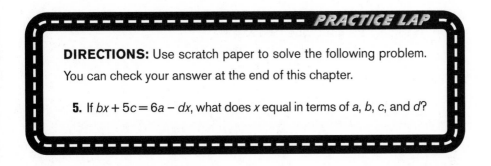

DIRECTIONS: Use scratch paper to solve the following problem. You can check your answer at the end of this chapter.

5. If $bx + 5c = 6a - dx$, what does x equal in terms of a, b, c, and d?

FRACTIONS WITH VARIABLES

You can work with fractions that contain variables in the same way as you would work with fractions without variables.

Example

Write $\frac{x}{6} - \frac{x}{12}$ as a single fraction.

First determine the LCD of 6 and 12: The LCD is 12. Then convert each fraction into an equivalent fraction with 12 as the denominator:

$$\frac{x}{6} - \frac{x}{12} = \frac{x \times 2}{6 \times 2} - \frac{x}{12} = \frac{2x}{12} - \frac{x}{12}$$

Then simplify:

$$\frac{2x}{12} - \frac{x}{12} = \frac{x}{12}$$

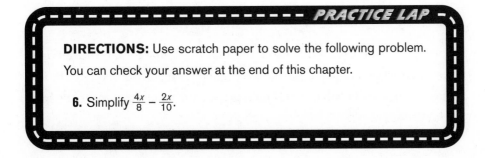

DIRECTIONS: Use scratch paper to solve the following problem. You can check your answer at the end of this chapter.

6. Simplify $\frac{4x}{8} - \frac{2x}{10}$.

RECIPROCAL RULES

There are special rules for the sum and difference of reciprocals. The following formulas can be memorized to save time when answering questions about reciprocals:

- ➡ If x and y are not 0, then $\frac{1}{x} + y = x + \frac{y}{xy}$
- ➡ If x and y are not 0, then $\frac{1}{x} - \frac{1}{y} = y - \frac{x}{xy}$

FINDING ROOTS

The roots of an equation are the values that make the equation true. For example, what are the roots of $x^3 - 9x^2 - 10x = 0$?

First, factor out the variable x: $x(x^2 - 9x - 10)$. Then, factor again: $x^2 - 9x - 10 = (x - 10)(x + 1)$, and $x^3 - 9x^2 - 10x = x(x - 10)(x + 1)$. Set each factor equal to 0 and solve for x: $x = 0$; $x - 10 = 0$, $x = 10$; $x + 1 = 0$, $x = -1$. The roots of this equation are 0, –1, and 10.

- - - **PRACTICE LAP** -

DIRECTIONS: Use scratch paper to solve the following problem. You can check your answer at the end of this chapter.

7. What is a root of $x(x - 1)(x + 1) = 27 - x$?

UNDEFINED EXPRESSIONS

A fraction is undefined when its denominator is equal to 0. If the denominator of a fraction is a polynomial, factor it and set the factors equal to 0. The values that make the polynomial equal to 0 are the values that make the fraction undefined.

For what values of x is the fraction $\frac{x + 49}{x^3 + 7x^2}$ undefined?

Factor the denominator and set each factor equal to 0 to find the values of x that make the fraction undefined: $x^3 + 7x^2 = x^2(x + 7)$, $x^2 = 0$, $x = 0$; $x + 7 = 0$, $x = -7$. The fraction is undefined when x equals 0 or -7.

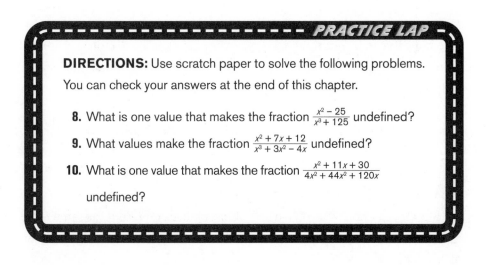

PRACTICE LAP

DIRECTIONS: Use scratch paper to solve the following problems. You can check your answers at the end of this chapter.

8. What is one value that makes the fraction $\frac{x^2 - 25}{x^3 + 125}$ undefined?

9. What values make the fraction $\frac{x^2 + 7x + 12}{x^3 + 3x^2 - 4x}$ undefined?

10. What is one value that makes the fraction $\frac{x^2 + 11x + 30}{4x^3 + 44x^2 + 120x}$

undefined?

RADICALS

In order to simplify some expressions and solve some equations, you will need to find the square or cube root of a number or variable. The **radical** symbol, $\sqrt{\ }$, signifies the root of a value. The square root, or second root, of x is equal to $\sqrt{x}$, or $\sqrt[2]{x}$. If there is no root number given, it is assumed that the radical symbol represents the square root of the number. The number under the radical symbol is called the **radicand.**

Adding and Subtracting Radicals

Two radicals can be added or subtracted if they have the same radicand. To add two radicals with the same radicand, add the coefficients of the radicals and keep the radicand the same.

$$2\sqrt{2} + 3\sqrt{2} = 5\sqrt{2}$$

To subtract two radicals with the same radicand, subtract the coefficient of the second radical from the coefficient of the first radical and keep the radicand the same.

$$6\sqrt{5} - 4\sqrt{5} = 2\sqrt{5}$$

The expressions $\sqrt{3} + \sqrt{2}$ and $\sqrt{3} - \sqrt{2}$ cannot be simplified any further, because these radicals have different radicands.

Multiplying Radicals

Two radicals can be multiplied whether they have the same radicand or not. To multiply two radicals, multiply the coefficients of the radicals and multiply the radicands.

$$(4\sqrt{6})(3\sqrt{7}) = 12\sqrt{42}, \text{ because } (4)(3) = 12 \text{ and } (\sqrt{6})(\sqrt{7}) = \sqrt{42}.$$

If two radicals of the same root with the same radicand are multiplied, the product is equal to the value of the radicand alone.

Here's an example: $(\sqrt{6})(\sqrt{6}) = 6$. Both radicals represent the same root, the square root, and both radicals have the same radicand, 6, so the product of $\sqrt{6}$ and $\sqrt{6}$ is 6.

Dividing Radicals

Two radicals can be divided whether they have the same radicand or not. To divide two radicals, divide the coefficients of the radicals and divide the radicands.

$$\frac{10\sqrt{15}}{2\sqrt{3}} = 5\sqrt{5}, \text{ because } \frac{10}{2} = 5 \text{ and } \frac{\sqrt{15}}{\sqrt{3}} = \sqrt{5}.$$

Any radical divided by itself is equal to 1: $\frac{\sqrt{3}}{\sqrt{3}} = 1$.

Simplifying a Single Radical

To simplify a radical such as $\sqrt{64}$, find the square root of 64. Look for a number that, when multiplied by itself, equals 64. Because $(8)(8) = 64$, the square root of 64 is 8: $\sqrt{64} = 8$; $\sqrt{64}$ is expressed as 8, not -8. The equation $x^2 = 64$ has two solutions, because both 8 and -8 square to 64, but the square root of a positive number is always its principal root (a positive number) when one exists.

However, most radicals cannot be simplified so easily. Many whole numbers and fractions do not have roots that are also whole numbers or fractions. You can simplify the original radical, but you will still have a radical in your answer.

To simplify a single radical, such as $\sqrt{32}$, find two factors of the radicand, one of which is a perfect square: $\sqrt{32} = (\sqrt{16})(\sqrt{2})$. Notice that $\sqrt{16}$ is a perfect square; the positive square root of 16 is 4. So, $\sqrt{32} = (\sqrt{16})(\sqrt{2}) = 4\sqrt{2}$.

RATIONALIZING DENOMINATORS OF FRACTIONS

An expression is not in simplest form if there is a radical in the denominator of a fraction. For example, the fraction $\frac{4}{\sqrt{3}}$ is not in simplest form. Multiply the top and bottom of the fraction by the radical in the denominator: Multiply $\frac{4}{\sqrt{3}}$ by $\frac{\sqrt{3}}{\sqrt{3}}$. Because $\frac{\sqrt{3}}{\sqrt{3}} = 1$, this will not change the value of the fraction. Because any radical multiplied by itself is equal to the radicand, $(\sqrt{3})(\sqrt{3}) = 3$; $(4)(\sqrt{3}) = 4\sqrt{3}$, so the fraction $\frac{4}{\sqrt{3}}$ in simplest form is $\frac{4\sqrt{3}}{3}$.

SOLVING EQUATIONS WITH RADICALS

Use the properties of adding, subtracting, multiplying, dividing, and simplifying radicals to help you solve equations with radicals. To remove a radical symbol from one side of an equation, you can raise both sides of the equation to a power. Remove a square root symbol from an equation by squaring both sides of the equation. Remove a cube root symbol from an equation by cubing both sides of the equation.

If $\sqrt{x} = 6$, what is the value of x?

To remove the radical symbol from the left side of the equation, square both sides of the equation. In other words, raise both sides of the equation to the power that is equal to the root of the radical. To remove a square root, or second root, raise both sides of the equation to the second power. To remove a cube root, or third root, raise both sides of the equation to the third power.

$$\sqrt{x} = 6, (\sqrt{x})^2 = (6)^2, x = 36$$
$$\sqrt[3]{x} = 3, (\sqrt[3]{x})^3 = (3)^3, x = 27$$

EXPONENTS

When a value, or base, is raised to a power, that power is the **exponent** of the base. The exponent of the term 4^2 is 2, and the base of the term is 4. The exponent is equal to the number of times a base is multiplied by itself: $4^2 = (4)(4)$; $2^6 = (2)(2)(2)(2)(2)(2)$.

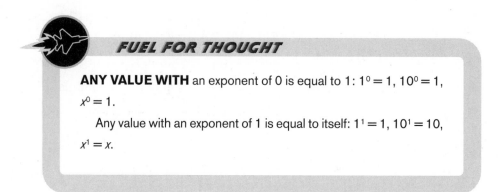

FUEL FOR THOUGHT

ANY VALUE WITH an exponent of 0 is equal to 1: $1^0 = 1$, $10^0 = 1$, $x^0 = 1$.

Any value with an exponent of 1 is equal to itself: $1^1 = 1$, $10^1 = 10$, $x^1 = x$.

Fractional Exponents

An exponent can also be a fraction. The numerator of the fraction is the power to which the base is being raised. The denominator of the fraction is the root of the base that must be taken. For example, the square root of a number can be represented as $x^{\frac{1}{2}}$, which means that x must be raised to the first power ($x^1 = x$) and then the second, or square, root must be taken: $x^{\frac{1}{2}} = \sqrt[2]{x^1} = \sqrt{x}$.

$$4^{\frac{3}{2}} = (\sqrt{4})^3 = 2^3 = 8$$

It does not matter if you find the root (represented by the denominator) first, and then raise the result to the power (represented by the numerator), or if you find the power first and then take the root.

$$4^{\frac{3}{2}} = \sqrt{(4)^3} = 64, \sqrt{64} = 8$$

Negative Exponents

A base raised to a negative exponent is equal to the reciprocal of the base raised to the positive value of that exponent.

$$3^{-3} = \frac{1}{3^3}$$

$$x^{-2} = \frac{1}{x^2}$$

Multiplying and Dividing Terms with Exponents

To multiply two terms with common bases, multiply the coefficients of the bases and add the exponents of the bases.

$$(3x^2)(7x^4) = 21x^6$$

$$(2x^{-5})(2x^3) = 4x^{-2}, \text{ or } \frac{4}{x^2}$$

$$(x^c)(x^d) = x^{c+d}$$

To divide two terms with common bases, divide the coefficients of the bases and subtract the exponents of the bases.

$$\frac{27x^5}{9x} = 3x^4$$

$$\frac{2x^3}{8x^4} = \frac{x^{-1}}{4}, \text{ or } \frac{1}{(4x)}$$

$$\frac{x^c}{x^d} = x^{c-d}$$

Raising a Term with an Exponent to Another Exponent

When a term with an exponent is raised to another exponent, keep the base of the term and multiply the exponents.

$$(x^3)^3 = x^9$$

$$(x^c)^d = x^{cd}$$

If the term that is being raised to an exponent has a coefficient, be sure to raise the coefficient to the exponent as well.

$$(3x^2)^3 = 27x^6$$

$$(cx^3)^4 = c^4x^{12}$$

PRACTICE LAP

DIRECTIONS: Use scratch paper to solve the following problems. You can check your answers at the end of this chapter.

11. $\sqrt{32x^2} =$

12. $\dfrac{\sqrt[3]{27y^3}}{\sqrt{27y^2}} =$

13. $\left(\dfrac{\sqrt{m^3}}{n^5}\right)^{-2} =$

14. What is the value of $(x^{-y})(2x^v)(3y^x)$ if $x = 2$ and $y = -2$?

ANSWERS

1. To find the quotient:

   $\dfrac{18x^8y^5}{24x^3y^4}$

   Divide the coefficients and subtract the exponents.

   $\dfrac{3x^{8-3}y^{5-4}}{4}$

   $\dfrac{3x^5y^1}{4}$

   $= \dfrac{3x^5y}{4}$

2. To find the product, follow the FOIL method:

   $(7x + 3)(5x - 2)$

   F: $7x$ and $5x$ are the **first** pair of terms.

   O: $7x$ and -2 are the **outer** pair of terms.

   I: 3 and $5x$ are the **inner** pair of terms.

   L: 3 and -2 are the **last** pair of terms.

   Now multiply according to FOIL:

   $(7x)(5x) + (7x)(-2) + (3)(5x) + (3)(-2) = 35x^2 - 14x + 15x - 6$

   Now combine like terms:

   $35x^2 + x - 6$

3. The rule $a^2 - b^2 = (a + b)(a - b)$ applies only to the difference between perfect squares, and 27, 40, 48, and 72 are not perfect squares. Choice e, 64, is a perfect square, so $x^2 - 64$ can be factored as $(x + 8)(x - 8)$.

4. To find the factors, determine a common factor for each term of $56a^5 + 21a$. Both coefficients (56 and 21) can be divided by 7 and both variables can

be divided by a. Therefore, a common factor is $7a$. Now, to find the second factor, divide the polynomial by the first factor:

$$\frac{56a^5 + 21a}{7a}$$

$$\frac{8a^5 + 3a^1}{a^1}$$

Subtract exponents when dividing.

$8a^{5-1} + 3a^{1-1}$

$8a^4 + 3a^0$

A base with an exponent of $0 = 1$.

$8a^4 + 3(1)$

$8a^4 + 3$

Therefore, the factors of $56a^5 + 21a$ are $7a(8a^4 + 3)$.

5. Use factoring to isolate x:

$bx + 5c = 6a - dx$

First isolate the x terms on the same side.

$bx + 5c + dx = 6a - dx + dx$

$bx + 5c + dx = 6a$

$bx + 5c + dx - 5c = 6a - 5c$

$bx + dx = 6a - 5c$

Now factor out the common x term.

$x(b + d) = 6a - 5c$

Now divide to isolate x.

$$\frac{x(b + d)}{b + d} = \frac{6a - 5c}{b + d}$$

$$x = \frac{6a - 5c}{b + d}$$

6. To simplify the expression, first determine the LCD of 8 and 10: The LCD is 40. Then convert each fraction into an equivalent fraction with 40 as the denominator:

$$\frac{4x}{8} - \frac{2x}{10} = \frac{4x \times 5}{8 \times 5} - \frac{2x \times 4}{10 \times 4} = \frac{20x}{40} - \frac{8x}{40}$$

Then simplify and reduce:

$$\frac{20x}{40} - \frac{8x}{40} = \frac{12x}{40} = \frac{3x}{10}$$

7. The root is 3. First, multiply the terms on the left side of the equation. $x(x - 1) = x^2 - x$, $(x^2 - x)(x + 1) = x^3 + x^2 - x^2 - x = x^3 - x$. Therefore, $x^3 - x = 27 - x$. Add x to both sides of the equation: $x^3 - x + x = 27 - x + x$, $x^3 = 27$. The cube root of 27 is 3, so the root, or solution, of $x(x - 1)$ $(x + 1) = 27 - x$ is $x = 3$.

8. A fraction is undefined when its denominator is equal to 0. Set the denominator equal to 0 and solve for x: $x^3 + 125 = 0$, $x^3 = -125$, $x = -5$.

9. A fraction is undefined when its denominator is equal to 0. Factor the polynomial in the denominator and set each factor equal to 0 to find the values that make the fraction undefined: $x^3 + 3x^2 - 4x = x(x + 4)(x - 1)$; $x = 0$; $x + 4 = 0$, $x = -4$; $x - 1 = 0$, $x = 1$. The fraction is undefined when x is equal to -4, 0, or 1.

10. A fraction is undefined when its denominator is equal to 0. Factor the polynomial in the denominator and set each factor equal to 0 to find the values that make the fraction undefined: $4x^3 + 44x^2 + 120x = 4x(x^2 + 11x + 30) = 4x(x + 5)(x + 6)$; $4x = 0$, $x = 0$; $x + 5 = 0$, $x = -5$; $x + 6 = 0$, $x = -6$. The fraction is undefined when x is equal to -6, -5, or 0.

11. Find the square root of the coefficient and the variable: $\sqrt{(32x^2)} = \sqrt{32}\sqrt{(x^2)} = x\sqrt{32}$. Next, factor $\sqrt{32}$ into two radicals, one of which is a perfect square: $\sqrt{32} = (\sqrt{16})(\sqrt{2}) = 4\sqrt{2}$. Therefore, $\sqrt{(32x^2)} = 4x\sqrt{2}$.

12. The cube root of $27y^3 = 3y$, because $(3y)(3y)(3y) = 27y^3$. Factor the denominator into two radicals: $\sqrt{(27y^2)} = (\sqrt{(9y^2)})(\sqrt{3})$. The square root of $9y^2 = 3y$, because $(3y)(3y) = 9y^2$. The expression is now equal to $\frac{3y}{3y\sqrt{3}}$. Cancel the $3y$ terms from the numerator and denominator, leaving $\frac{1}{\sqrt{3}}$. Simplify the fraction by multiplying the numerator and denominator by $\sqrt{3}$: $\left(\frac{1}{\sqrt{3}}\right)\left(\frac{\sqrt{3}}{\sqrt{3}}\right) = \frac{\sqrt{3}}{3}$.

13. A term with a negative exponent can be rewritten as the reciprocal of the term with a positive exponent: $\left(\sqrt{\frac{m^3}{n^5}}\right)^{-2} = \left(\frac{1}{\frac{\sqrt{m^3}}{n^5}}\right)^2$. Square the numerator and denominator: $(1)^2 = 1$, $\left(\sqrt{\frac{m^3}{n^5}}\right)^2 = \frac{m^3}{n^5}$. Therefore, $\left(\frac{1}{\frac{\sqrt{m^3}}{n^5}}\right)^2 = \left(\frac{1}{\frac{m^3}{n^5}}\right) = \frac{n^5}{m^3}$.

14. First, multiply the first two terms. Multiply the coefficients of the terms and add the exponents. Because $-y + y = 0$, $(x^{-y})(2x^y) = 2$, and $(x^{-y})(2x^y)(3y^x) = 2(3y^x) = 6y^x$. Substitute 2 for x and -2 for y: $6(-2)^2 = 6(4) = 24$.

The Three Qs— Quadratic Trinomials, Quadratic Equations, and the Quadratic Formula

A **quadratic trinomial** contains an x^2 term as well as an x term. For example, $x^2 - 6x + 8$ is a quadratic trinomial. You can factor quadratic trinomials by using the FOIL method in reverse.

Example

Let's factor $x^2 - 6x + 8$.

Start by looking at the last term in the trinomial: 8. Ask yourself, "What two integers, when multiplied together, have a product of positive 8?" Make a mental list of these integers:

1×8 -1×-8 2×4 -2×-4

Next look at the middle term of the trinomial: $-6x$. Choose the two factors from the list you just made that also add up to the coefficient -6:

-2 and -4

Now write the factors using -2 and -4:

$(x - 2)(x - 4)$

Use the FOIL method to double-check your answer:

$(x - 2)(x - 4) = x^2 - 6x + 8$

You can see that the answer is correct.

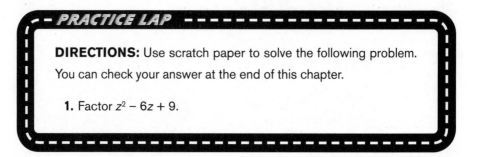

- - **PRACTICE LAP** -

DIRECTIONS: Use scratch paper to solve the following problem. You can check your answer at the end of this chapter.

1. Factor $z^2 - 6z + 9$.

QUADRATIC EQUATIONS

A **quadratic equation** is an equation that does not graph into a straight line. The graph will be a smooth curve. An equation is a quadratic equation if the highest exponent of the variable is 2. Here are some examples of quadratic equations:

$$x^2 + 6x + 10 = 0$$
$$6x^2 + 8x - 22 = 0$$

A quadratic equation can be written in the form: $ax^2 + bx + c = 0$. The a represents the number in front of the x^2 variable. The b represents the number in front of the x variable and c is the number. For instance, in the equation $2x^2 + 3x + 5 = 0$, the a is 2, the b is 3, and the c is 5. In the equation $4x^2 - 6x + 7 = 0$, the a is 4, the b is -6, and the c is 7. In the equation $5x^2 + 7 = 0$, the a is 5, the b is 0, and the c is 7. In the equation $8x^2 - 3x = 0$, the a is 8, the b is -3, and the c is 0. Is the equation $2x + 7 = 0$ a quadratic equation? No! The equation does not contain a variable with an exponent of 2. Therefore, it is not a quadratic equation.

Solving Quadratic Equations Using Factoring

Why is the equation $x^2 = 4$ a quadratic equation? It is a quadratic equation because the variable has an exponent of 2. To solve a quadratic equation, first make one side of the equation zero. Let's work with $x^2 = 4$.

Subtract 4 from both sides of the equation to make one side of the equation zero: $x^2 - 4 = 4 - 4$. Now, simplify $x^2 - 4 = 0$. The next step is to factor $x^2 - 4$. It can be factored as the difference of two squares: $(x - 2)(x + 2) = 0$.

If $ab = 0$, you know that either a or b or both factors have to be zero because a times $b = 0$. This is called the **zero product property**, and it says that if the product of two numbers is zero, then one or both of the numbers have to be zero. You can use this idea to help solve quadratic equations with the factoring method.

Use the zero product property, and set each factor equal to zero: $(x - 2) = 0$ and $(x + 2) = 0$.

When you use the zero product property, you get linear equations that you already know how to solve.

Solve the equation:	$x - 2 = 0$
Add 2 to both sides of the equation.	$x - 2 + 2 = 0 + 2$
Now, simplify:	$x = 2$
Solve the equation:	$x + 2 = 0$
Subtract 2 from both sides of the equation.	$x + 2 - 2 = 0 - 2$
Simplify:	$x = -2$

You got two values for x. The two solutions for x are 2 and –2. All quadratic equations have two solutions. The exponent 2 in the equation tells you that the equation is quadratic, and it also tells you that you will have two answers.

INSIDE TRACK

WHEN BOTH YOUR solutions are the same number, this is called a **double root**. You will get a double root when both factors are the same.

Before you can factor an expression, the expression must be arranged in descending order. An expression is in descending order when you start with the largest exponent and descend to the smallest, as shown in this example: $2x^2 + 5x + 6 = 0$.

All quadratic equations have two solutions. The exponent of 2 in the equation tells you to expect two answers.

Example

$x^2 - 3x - 4 = 0$

Factor the trinomial $x^2 - 3x - 4$.	$(x - 4)(x + 1) = 0$
Set each factor equal to zero.	$x - 4 = 0$ and $x + 1 = 0$
Solve the equation.	$x - 4 = 0$
Add 4 to both sides of the equation.	$x - 4 + 4 = 0 + 4$
Simplify.	$x = 4$
Solve the equation.	$x + 1 = 0$
Subtract 1 from both sides of the equation.	$x + 1 - 1 = 0 - 1$
Simplify.	$x = -1$

The two solutions for the quadratic equation are 4 and –1.

INSIDE TRACK

WHEN YOU HAVE an equation in factor form, disregard any factor that is a number and contains no variables. For example, in $4(x - 5)$ $(x + 5) = 0$, disregard the 4. It will have no effect on your two solutions.

PRACTICE LAP

DIRECTIONS: Use scratch paper to solve the following problems. You can check your answers at the end of this chapter.

2. Solve $4x^2 = 100$.

3. If $(x - 8)(x + 5) = 0$, what are the two possible values of x?

Solving Quadratic Equations by Using the Zero Product Rule

If a quadratic equation is not equal to zero, rewrite it so that you can solve it using the zero product rule.

Example

If you need to solve $x^2 - 11x = 12$, subtract 12 from both sides:

$x^2 - 11x - 12 = 12 - 12$

$x^2 - 11x - 12 = 0$

Now this quadratic equation can be solved using the zero product rule.

A quadratic equation must be factored before using the zero product rule to solve it.

Example

To solve $x^2 + 9x = 0$, first factor it:

$x(x + 9) = 0$

Now you can solve it.

Either $x = 0$ or $x + 9 = 0$.

Therefore, possible solutions are $x = 0$ and $x = -9$.

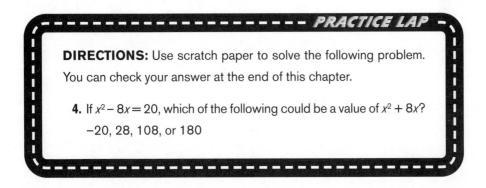

PRACTICE LAP

DIRECTIONS: Use scratch paper to solve the following problem. You can check your answer at the end of this chapter.

4. If $x^2 - 8x = 20$, which of the following could be a value of $x^2 + 8x$?

   −20, 28, 108, or 180

GRAPHS OF QUADRATIC EQUATIONS

The (x,y) solutions to quadratic equations can be plotted on a graph. These graphs are called **parabolas**. Typically you will be presented with parabolas given by equations in the form of $y = ax^2 + bx + c$.

Notice that the equation $y = x^2$ conforms to this formula—both b and c are zero.

$$y = (1)x^2 + (0)x + (0) \text{ is equivalent to } y = x^2$$

The value of a cannot equal zero, however.

If a is greater than zero, the parabola will open upward. If a is less than zero, the parabola will open downward.

The x-coordinate of the turning point, or vertex, of the parabola is given by:

$$x = \frac{-b}{2a}$$

You can use this x-value in the original formula and solve for y (the y-coordinate of the turning point).

There will also be a line of symmetry given by:

$$x = \frac{-b}{2a}$$

For the graph $y = x^2$, $x = \frac{-b}{2a} = 0$. The line of symmetry is $x = 0$. The y-coordinate of the vertex is at located at $y = x^2 = 0^2 = 0$, so the vertex is at $(0,0)$. Technically a parabola could also be given by the formula $x = ay^2 + by + c$.

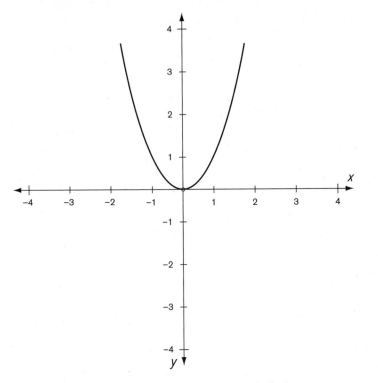

The graph of the equation $y = x^2$ is a parabola.

Because the x-value is squared, the positive values of x yield the same y-values as the negative values of x. The graph of $y = x^2$ has its vertex at the point (0,0). The vertex of a parabola is the turning point of the parabola. It is either the minimum or maximum y-value of the graph. The graph of $y = x^2$ has its minimum at (0,0). There are no y-values less than 0 on the graph.

The graph of $y = x^2$ can be translated around the coordinate plane. While the parabola $y = x^2$ has its vertex at (0,0), the parabola $y = x^2 - 1$ has its vertex at (0,−1). After the x term is squared, the graph is shifted down one unit. A parabola of the form $y = x^2 - c$ has its vertex at (0,−c) and a parabola of the form $y = x^2 + c$ has its vertex at (0,c).

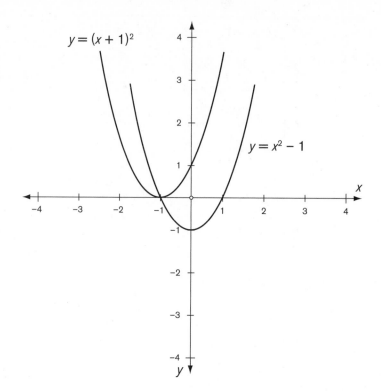

The parabola $y = (x + 1)^2$ has its vertex at $(-1,0)$. The x-value is increased before it is squared. The minimum value of the parabola is when $y = 0$ (because $y = (x + 1)^2$ can never have a negative value). The expression $(x + 1)^2$ is equal to 0 when $x = -1$. A parabola of the form $y = (x - c)^2$ has its vertex at $(c,0)$ and a parabola of the form $y = (x + c)^2$ has its vertex at $(0,-c)$.

What are the coordinates of the vertex of the parabola formed by the equation $y = (x - 2)^2 + 3$?

To find the x-value of the vertex, set $(x - 2)$ equal to 0: $x - 2 = 0$, $x = 2$. The y-value of the vertex of the parabola is equal to the constant that is added to or subtracted from the x squared term. The y-value of the vertex is 3, making the coordinates of the vertex of the parabola $(2,3)$.

If parabolas with the formula $y = x^2 + bx + c$ open upward or downward, how do you think parabolas given by the formula $x = ay^2 + by + c$ appear?

It is important to be able to look at an equation and understand what its graph will look like. You must be able to determine what calculation to perform on each x-value to produce its corresponding y-value.

For example, here is the graph of $y = x^2$.

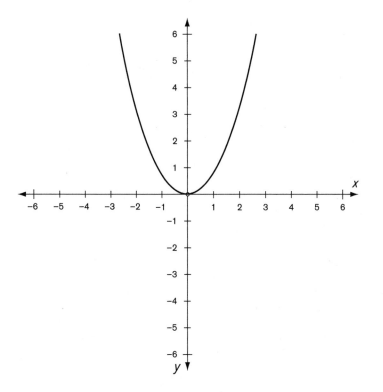

The equation $y = x^2$ tells you that for every x-value, you must square the x-value to find its corresponding y-value. Let's explore the graph with a few x-coordinates:

An x-value of 1 produces what y-value? Plug $x = 1$ into $y = x^2$. When $x = 1$, $y = 1^2$, so $y = 1$. So, you know a coordinate in the graph of $y = x^2$ is (1,1).

INSIDE TRACK

SOLVING THE FORMULA of a parabola for x tells you the x intercept (or intercepts) of the parabola—that is, where the parabola crosses the x-axis. If you get two real values for x, the parabola crosses the x-axis at two points. If you get one real root, then that value is the vertex. If both roots are complex, then the parabola never crosses the x-axis.

An x-value of 2 produces what y-value? Plug $x = 2$ into $y = x^2$. When $x = 2$, $y = 2^2$, so $y = 4$. Therefore, you know a coordinate in the graph of $y = x^2$ is (2,4).

An x-value of 3 produces what y-value? Plug $x = 3$ into $y = x^2$. When $x = 3$, $y = 3^2$, so $y = 9$. That determines that a coordinate in the graph of $y = x^2$ is (3,9).

Okay, now what if you are asked to compare the graph of $y = x^2$ with the graph of $y = (x - 1)^2$? Let's compare what happens when you plug numbers (x-values) into $y = (x - 1)^2$ with what happens when you plug numbers (x-values) into $y = x^2$:

$y = x^2$	$y = (x - 1)^2$
If $x = 1, y = 1.$	If $x = 1, y = 0.$
If $x = 2, y = 4.$	If $x = 2, y = 1.$
If $x = 3, y = 9.$	If $x = 3, y = 4.$
If $x = 4, y = 16.$	If $x = 4, y = 9.$

The two equations have the same y-values, but they match up with different x-values because $y = (x - 1)^2$ subtracts 1 before squaring the x-value. As a result, the graph of $y = (x - 1)^2$ looks identical to the graph of $y = x^2$ except that the base is shifted to the right (on the x-axis) by 1:

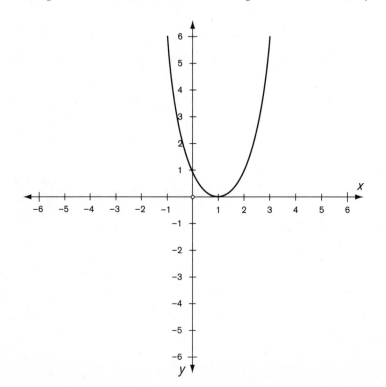

How would the graph of $y = x^2$ compare with the graph of $y = x^2 - 1$?

In order to find a y-value with $y = x^2$, you square the x-value. In order to find a y-value with $y = x^2 - 1$, square the x-value and then subtract 1. This means the graph of $y = x^2 - 1$ looks identical to the graph of $y = x^2$ except that the base is shifted down (on the y-axis) by 1:

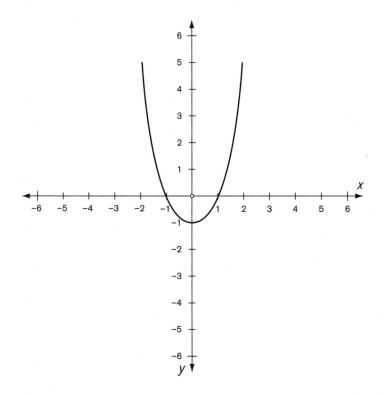

DIRECTIONS: Use scratch paper and graph paper to solve the following problem. You can check your answer at the end of this chapter.

5.

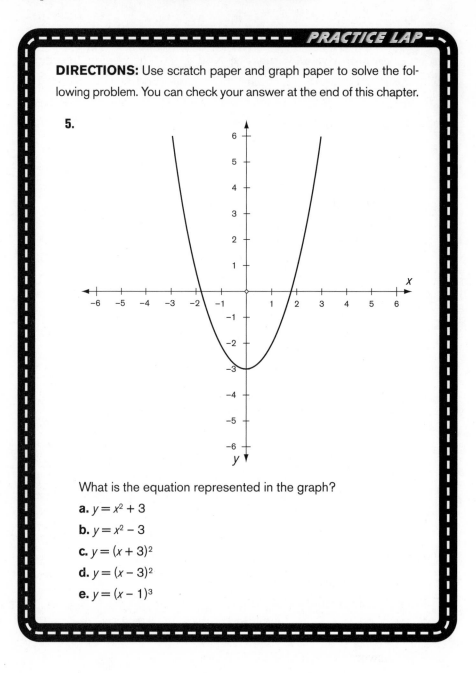

What is the equation represented in the graph?

a. $y = x^2 + 3$

b. $y = x^2 - 3$

c. $y = (x + 3)^2$

d. $y = (x - 3)^2$

e. $y = (x - 1)^3$

WORD PROBLEMS AREN'T A PROBLEM

You can easily solve the word problems using quadratic equations. Let's look carefully at an example.

Example

You have a patio that is 8 ft. by 10 ft. You want to increase the size of the patio to 168 square ft. by adding the same length to both sides of the patio. Let x = the length you will add to each side of the patio. You find the area of a rectangle by multiplying the length times the width. The new area of the patio will be 168 square ft.

$(x + 8)(x + 10) = 168$

FOIL the factors $(x + 8)(x + 10)$.	$x^2 + 10x + 8x + 80 = 168$
Simplify.	$x^2 + 18x + 80 = 168$
Subtract 168 from both sides of the equation.	$x^2 + 18x + 80 - 168 = 168 - 168$
Simplify both sides of the equation.	$x^2 + 18x - 88 = 0$
Factor.	$(x + 22)(x - 4) = 0$
Set each factor equal to zero.	$x + 22 = 0$ and $x - 4 = 0$
Solve the equation.	$x + 22 = 0$
Subtract 22 from both sides of the equation.	$x + 22 - 22 = 0 - 22$
Simplify both sides of the equation.	$x = -22$
Solve the equation.	$x - 4 = 0$
Add 4 to both sides of the equation.	$x - 4 + 4 = 0 + 4$
Simplify both sides of the equation.	$x = 4$

Because this is a quadratic equation, you can expect two answers. The answers are 4 and −22. However, −22 is not a reasonable answer. You cannot have a negative length. Therefore, the only answer is 4.

To check your calculations, review the original dimensions of the patio—8 ft. by 10 ft. If you were to add 4 to each side, the new dimensions would be 12 ft. by 14 ft. When you multiply 12 times 14, you get 168 square ft., which is the new area you wanted.

WHAT IS THE QUADRATIC FORMULA?

You can use the quadratic formula to solve quadratic equations. You might be asking yourself, "Why do I need to learn another method for solving quadratic equations when I already know how to solve them by using factoring?" Well, not all quadratic equations can be solved using factoring. You use the factoring method because it is faster and easier, but it will not always work. However, the quadratic formula will always work.

   The quadratic formula is a formula that allows you to solve any quadratic equation—no matter how simple or difficult. If the equation is written in the form $ax^2 + bx + c = 0$, then the two solutions for x will be $x = \frac{-b \pm \sqrt{b^2 - 4ac}}{2a}$. It is the $\pm$ in the formula that gives us the two answers: one with + in that spot, and one with −. The formula contains a radical, which is one of the reasons you studied radicals in the previous lesson. To use the formula, you substitute the values of a, b, and c into the formula and then carry out the calculations.

Example
   $3x^2 - x - 2 = 0$

Determine a, b, and c. $\qquad$ $a = 3$, $b = -1$, and $c = -2$

Take the quadratic formula. $\qquad$ $\frac{-b \pm \sqrt{b^2 - 4ac}}{2a}$

Substitute in the values of a, b, and c. $\qquad$ $\frac{-[-1] \pm \sqrt{-1^2 - 4(3)(-2)}}{2 \cdot 3}$

Simplify. $\qquad$ $\frac{1 \pm \sqrt{1 - [-24]}}{6}$

Simplify more. $\qquad$ $\frac{1 \pm \sqrt{25}}{6}$

Take the square root of 25. $\qquad$ $\frac{1 \pm 5}{6}$

The solutions are 1 and $-\frac{2}{3}$. $\qquad$ $\frac{1 + 5}{6} = \frac{6}{6} = 1$ and

$\qquad\qquad\qquad\qquad\qquad\qquad$ $\frac{1 - 5}{6} = -\frac{4}{6} = -\frac{2}{3}$

CAUTION!

TO USE THE quadratic formula, you need to know the a, b, and c of the equation. However, before you can determine what a, b, and c are, the equation must be in $ax^2 + bx + c = 0$ form. For example, the equation $5x^2 + 2x = 9$ must be transformed to $ax^2 + bx + c = 0$ form.

SOLVING QUADRATIC EQUATIONS THAT HAVE A RADICAL IN THE ANSWER

Some equations will have radicals in their answers. The strategy for solving these equations is the same as the equations you just completed.

Example

$$3m^2 - 3m = 1$$

Subtract 1 from both sides of the
equation. $\qquad$ $3m^2 - 3m - 1 = 1 - 1$

Simplify. $\qquad$ $3m^2 - 3m - 1 = 0$

Use the quadratic formula with
$a = 3$, $b = -3$, and $c = -1$. $\qquad$ $\frac{-b \pm \sqrt{b^2 - 4ac}}{2a}$

Substitute the values for a, b, and c. $\quad \frac{-[-3] \pm \sqrt{[-3]^2 - 4(3)(-1)}}{2 \cdot 3}$

Simplify. $\qquad$ $\frac{3 \pm \sqrt{9 - [-12]}}{6}$

Simplify. $\qquad$ $\frac{3 \pm \sqrt{21}}{6}$

The solution to the equation is $m = \frac{(3 \pm \sqrt{21})}{6}$ because one answer

is $m = \frac{(3 + \sqrt{21})}{6}$ and the other answer is $m = \frac{(3 - \sqrt{21})}{6}$.

ANSWERS

1. To find the factors, follow the FOIL method in reverse:

   $z^2 - 6z + 9$

   The product of the **last** pair of terms equals $+9$. There are a few possibilities for these terms: 3 and 3 (because $3 \times 3 = +9$), -3 and -3 (because $-3 \times -3 = +9$), 9 and 1 (because $9 \times 1 = +9$), -9 and -1 (because $-9 \times -1 = +9$).

   The sum of the products of the **outer** pair of terms and the **inner** pair of terms equals $-6z$. So we must choose the two last terms from the list of possibilities that would add up to -6. The only possibility is -3 and -3. Therefore, we know that the last terms are -3 and -3.

   The product of the **first** pair of terms equals z^2. The most likely two terms for the first pair is z and z because $z \times z = z^2$.

   Therefore, the factors are $(z - 3)(z - 3)$.

2. Make one side of the equation zero. Subtract 100 from both sides of the equation: $4x^2 - 100 = 100 - 100$. Now, simplify: $4x^2 - 100 = 0$. Factor out the greatest common factor: $4(x^2 - 25) = 0$. Factor using the difference of two squares: $4(x - 5)(x + 5) = 0$. Divide both sides of the equation by 4: $\frac{4(x - 5)(x + 5)}{4} = \frac{0}{4}$. Time to simplify again: $(x - 5)(x + 5) = 0$. Set each factor equal to zero: $x - 5 = 0$ and $x + 5 = 0$. Solve the equations: $x = 5$ and $x = -5$. The solutions for the quadratic equation are 5 and -5.

3. If $(x - 8)(x + 5) = 0$, then one (or both) of the factors must equal 0.

   $x - 8 = 0$ if $x = 8$ because $8 - 8 = 0$.

   $x + 5 = 0$ if $x = -5$ because $-5 + 5 = 0$.

   So, the two values of x that make $(x - 8)(x + 5) = 0$ are $x = 8$ and $x = -5$.

4. This question requires several steps to answer. First, you must determine the possible values of x considering that $x^2 - 8x = 20$. To find the possible x-values, rewrite $x^2 - 8x = 20$ as $x^2 - 8x - 20 = 0$, factor, and then use the zero product rule; $x^2 - 8x - 20 = 0$ is factored as $(x - 10)(x + 2)$. So, possible values of x are $x = 10$ and $x = -2$ because $10 - 10 = 0$ and $-2 + 2 = 0$.

   Now, to find possible values of $x^2 + 8x$, plug in the x-values. If $x = -2$, then $x^2 + 8x = (-2)^2 + (8)(-2) = 4 + (-16) = -12$. None of the answer choices are -12, so try $x = 10$. If $x = 10$, then $x^2 + 8x = 10^2 + (8)(10) = 100 + 80 = 180$.

5. This graph is identical to a graph of $y = x^2$ except it is moved down 3 so that it intersects the y-axis at -3 instead of 0. Each y-value is 3 less than the corresponding y-value in $y = x^2$, so its equation is therefore $y = x^2 - 3$ (choice **b**).

Sequences and Series

sequence is a series of terms in which each term in the series is generated using a rule. Each value in the sequence is a called a **term**. The rule of a sequence could be "each term is twice the previous term" or "each term is 4 more than the previous term."

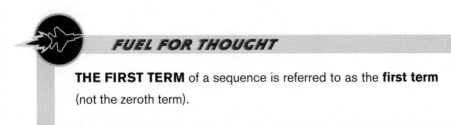

FUEL FOR THOUGHT

THE FIRST TERM of a sequence is referred to as the **first term** (not the zeroth term).

ARITHMETIC SEQUENCES

An **arithmetic sequence** is a series of terms in which the difference between any two consecutive terms in the sequence is always the same. For example, 2, 6, 10, 14, 18, . . . is an arithmetic sequence. The difference between any two consecutive terms is always +4.

Each term in the following sequence is five less than the previous term. What is the next term in the sequence?

38, 33, 28, 23, . . .

To find the next term in the sequence, take the last term given and subtract 5, because the rule of the sequence is "each term in the sequence is five less than the previous term." You know that 23 – 5 = 18, so the next term in the sequence is 18.

GEOMETRIC SEQUENCES

When analyzing a sequence, try to find the mathematical operation that you can perform to get the next number in the sequence. Let's try an example. Look carefully at the following sequence:

2, 4, 8, 16, 32, . . .

Notice that each successive term is found by multiplying the prior term by 2: $2 \times 2 = 4$, $4 \times 2 = 8$, and so on. Because each term is multiplied by a constant number (2), there is a constant ratio between the terms. Sequences that have a constant ratio between terms are called **geometric sequences**.

A geometric sequence is a series of terms in which the ratio between any two consecutive terms in the sequence is always the same. For example, 1, 3, 9, 27, 81, . . . is a geometric sequence. The ratio between any two consecutive terms is always 1:3—each term is three times the previous term.

Each term in the following sequence is six times the previous term. What is the value of x?

2, 12, x, 432, . . .

To find the value of x, the third term in the sequence, multiply the second term in the sequence, 12, by 6, because every term is six times the previous term: $(12)(6) = 72$. The third term in the sequence is 72. You can check your answer by multiplying 72 by 6: $(72)(6) = 432$, the fourth term in the sequence.

On an exam, you may be asked to determine a specific term in a sequence. Sometimes you will be asked for the 20th, 50th, or 100th term of a sequence. It would be unreasonable in many cases to evaluate that term, but you can represent that term with an expression.

Each term in the following sequence is four times the previous term. What is the hundredth term of the sequence?

3, 12, 48, 192, . . .

Write each term of the sequence as a product: 3 is equal to $4^0 \times 3$, 12 is equal to $4^1 \times 3$, 48 is equal to $4^2 \times 3$, and 192 is equal to $4^3 \times 3$. Each term in the sequence is equal to 4 raised to an exponent, multiplied by 3. For each term, the value of the exponent is one less than the position of the term in the sequence. The fourth term in the sequence, 192, is equal to 4 raised to one less than four (3), multiplied by 3. Therefore, the 100th term of the sequence is equal to 4 raised to one less than 100 (99), multiplied by 3. The hundredth term is equal to $4^{99} \times 3$.

Let's say you are asked to find the 30th term of a geometric sequence like 2, 4, 8, 16, 32, . . . You could answer such a question by writing out 30 terms of a sequence, but this is an inefficient method. It takes too much time. Let's determine the formula:

First, let's evaluate the terms.
2, 4, 8, 16, 32, . . .
Term 1 = 2
Term 2 = 4, which is 2×2
Term 3 = 8, which is $2 \times 2 \times 2$
Term 4 = 16, which is $2 \times 2 \times 2 \times 2$
You can also write out each term using exponents:
Term 1 = 2
Term 2 = 2×2^1

Term 3 = 2×2^2

Term 4 = 2×2^3

You can now write a formula:

Term n = $2 \times 2^{n-1}$

So, Term 30 = $2 \times 2^{30-1}$ = 2×2^{29}

The generic formula for a geometric sequence is Term n = $a_1 \times r^{n-1}$, where n is the term you are looking for, a_1 is the first term in the series, and r is the ratio that the sequence increases by. In the preceding example, n = 30 (the 30th term), a_1 = 2 (because 2 is the first term in the sequence), and r = 2 (because the sequence increases by a ratio of 2; each term is two times the previous term).

You can use the formula Term n = $a_1 \times r^{n-1}$ when determining a term in any geometric sequence.

COMBINATION SEQUENCES

Some sequences are defined by rules that are a combination of operations. The terms in these sequences do not differ by a constant value or ratio. For example, each number in a sequence could be generated by the rule "double the previous term and add one":

5, 11, 23, 47, 95, . . .

Each term in the following sequence is one less than four times the previous term. What is the next term in the sequence?

1, 3, 11, 43, . . .

Take the last given term in the sequence, 43, and apply the rule:

4(43) − 1 = 172 − 1 = 171

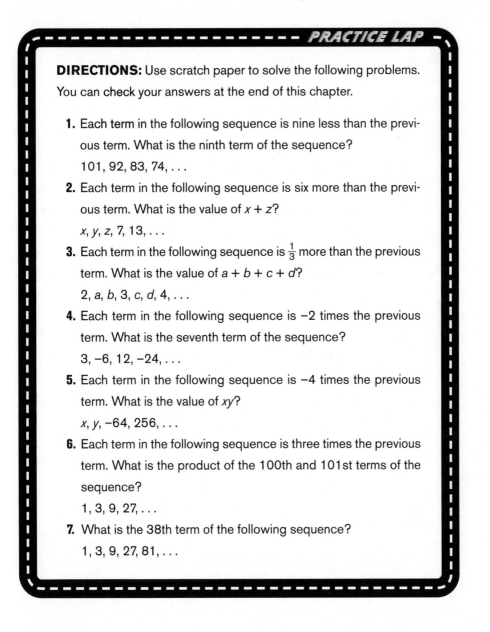

---- **PRACTICE LAP** --

DIRECTIONS: Use scratch paper to solve the following problems. You can check your answers at the end of this chapter.

1. Each term in the following sequence is nine less than the previous term. What is the ninth term of the sequence?

   101, 92, 83, 74, . . .

2. Each term in the following sequence is six more than the previous term. What is the value of $x + z$?

   $x, y, z, 7, 13, . . .$

3. Each term in the following sequence is $\frac{1}{3}$ more than the previous term. What is the value of $a + b + c + d$?

   $2, a, b, 3, c, d, 4, . . .$

4. Each term in the following sequence is −2 times the previous term. What is the seventh term of the sequence?

   3, −6, 12, −24, . . .

5. Each term in the following sequence is −4 times the previous term. What is the value of xy?

   $x, y, -64, 256, . . .$

6. Each term in the following sequence is three times the previous term. What is the product of the 100th and 101st terms of the sequence?

   1, 3, 9, 27, . . .

7. What is the 38th term of the following sequence?

   1, 3, 9, 27, 81, . . .

ANSWERS

1. The fourth term in the sequence is 74. You are looking for the ninth term, which is 5 terms after the fourth term. Because each term is nine less than the previous term, the ninth term will be 5(9) = 45 less than 74; 74 − 45 = 29. Because the number of terms is reasonable, you can check your answer by repeatedly subtracting 9: 74 − 9 = 65, 65 − 9 = 56, 56 − 9 = 47, 47 − 9 = 38, 38 − 9 = 29.

2. The term that follows z is 7. Because each term is 6 more than the previous term, z must be 6 less than 7. Therefore, $z = 7 - 6 = 1$. In the same way, y is 6 less than z and x is 6 less than y: $y = 1 - 6 = -5$ and $x = -5 - 6 = -11$. The sum of $x + z$ is equal to $-11 + 1 = -10$.

3. The first term in the sequence is 2. The next term in the sequence, a, is $\frac{1}{3}$ more than 2: $2\frac{1}{3}$. b is $\frac{1}{3}$ more than a, $2\frac{2}{3}$. c is $\frac{1}{3}$ more than 3: $3\frac{1}{3}$. d is $\frac{1}{3}$ more than c, $3\frac{2}{3}$. Add the values of a, b, c, and d: $2\frac{1}{3} + 2\frac{2}{3} + 3\frac{1}{3} + 3\frac{2}{3} = 12$.

4. The fourth term in the sequence is -24. You are looking for the seventh term, which is three terms after the fourth term. You must multiply by -2 three times, so the seventh term will be $(-2)^3 = -8$ times -24. $(-24)(-8) = 192$. Because the number of terms is reasonable, you can check your answer by repeatedly multiplying by -2: $(-24)(-2) = 48$, $(48)(-2) = -96$, $(-96)(-2) = 192$.

5. Because each term in the sequence is -4 times the previous term, y is equal to $\frac{-64}{-4} = 16$, and $x = \frac{16}{-4} = -4$. Therefore, $xy = (16)(-4) = -64$.

6. Every term in the sequence is 3 raised to a power. The first term, 1, is 3^0. The second term, 3, is 3^1. The value of the exponent is one less than the position of the term in the sequence. The 100th term of the sequence is equal to $3^{100-1} = 3^{99}$, and the 101st term in the sequence is equal to $3^{99+1} = 3^{100}$. To multiply two terms with common bases, add the exponents of the terms: $(3^{99})(3^{100}) = 3^{199}$.

7. 1, 3, 9, 27, 81, . . . is a geometric sequence. There is a constant ratio between terms: Each term is three times the previous term. You can use the formula Term $n = a_1 \times r^{n-1}$ to determine the nth term of this geometric sequence.

First determine the values of n, a_1, and r:

$n = 38$ (because you are looking for the 38th term)

$a_1 = 1$ (because the first number in the sequence is 1)

$r = 3$ (because the sequence increases by a ratio of 3; each term is three times the previous term)

Now solve:

Term $n = a_1 \times r^{n-1}$

Term $38 = 1 \times 3^{38-1}$

Term $38 = 1 \times 3^{37}$

10 Posttest

If you have completed all the chapters in this book, then you are ready to take the posttest to measure your progress. The posttest has 50 questions covering the topics you studied in this book. Although the format of the posttest is similar to that of the pretest, the questions are different.

Take as much time as you need to complete the posttest. When you are finished, check your answers with the answer key at the end of the posttest. Along with each answer is a number that tells you which chapter of this book teaches you the math skills needed for that question. Once you know your score on the posttest, compare the results with the pretest. If you scored better on the posttest than on the pretest, congratulations! You have profited from the hard work. At this point, you should look at the questions you missed, if any. Do you know why you missed the question, or do you need to go back to the chapter and review the concept?

If your score on the posttest doesn't show much improvement, take a second look at the questions you missed. Did you miss a question because of an error you made? If you can figure out why you missed the question, then you understand the concept and just need to concentrate more on accuracy when taking a test. If you missed a question because you did not know how to work a problem, go back to the chapter and spend time working that type of problem. Take time to understand algebra thoroughly. You need a solid foundation in algebra in order to succeed on tests or progress to a higher level of algebra. Whatever your score on this posttest, keep this book for review and future reference.

1. Given the parabola $y = 2x^2 - 8x + 1$, what is its vertex?

2. What is the vertex of the parabola with equation $y = 2x^2 + 16x$?

3. What is the vertex of the parabola with equation $y = 5x^2$?

4. Each term in the following sequence is nine more than $\frac{1}{3}$ the previous term. What is the value of $y - x$?
   81, 36, x, y, . . .

5. Each term in the following sequence is 20 less than five times the previous term. What is the value of $x + y$?
   x, 0, y, –120, . . .

6. Each term in the following sequence is two less than $\frac{1}{2}$ the previous term. What term of the sequence will be the first term to be a negative number?
   256, 126, 61, 28.5, . . .

7. Each term in the following sequence is 16 more than -4 times the previous term. What is the value of $x + y$?
   x, y, –80, 336, –1,328, . . .

8. Solve for c: $5\sqrt{c} + 15 = 35$

9. If $A = \begin{bmatrix} 3 & -1 & 2 \\ 0 & 2 & -4 \end{bmatrix}$, find $-A$.

10. $\dfrac{(\sqrt{(a^2b)})(\sqrt{(ab^2)})}{\sqrt{ab}} =$

11. $\left(\dfrac{(ab)^3}{b} \right)^4 =$

12. $((4g^2)^3(g^4))^{\frac{1}{2}} =$

13. If $a^{\frac{2}{3}} = 6$, then $a^{\frac{4}{3}} =$

14. If $(\sqrt{p})^4 = q^{-2}$, and $q = -\frac{1}{3}$, what is one possible value of p?

15. What is the value of $(a\sqrt{b})^{-ab}$ if $a = \frac{1}{3}$ and $b = 9$?

16. If $n = 20$, what is the value of $\left(\frac{\sqrt{n+5}}{\sqrt{n}}\right) \cdot \left(\frac{n}{2}\sqrt{5}\right)$?

17. If a is positive, and $a^2 = b = 4$, what is the value of $\left(\frac{b\sqrt{b}}{a^4}\right)^a$?

18. Graph the following inequality:

$x - y > 5$

19. Graph the following inequality:

$3x - y < 2$

20. Graph $y = 3x - 2$.

21. Use scratch paper and graph paper to solve the following system of inequalities.

$6x + 3y < 9$
$-x + y \le -4$

22. Use scratch paper and graph paper to solve the following system of inequalities.

$y \ge 6$
$x + y < 3$

23. Graph the following inequality:

$x \le -2$

24. Graph the following inequality:

$y \ge 0$

25. Which coordinate pair is a solution to the inequality $12 - 3y > 6x + 3$?
   a. (1,1)
   b. (2,1)
   c. (1,2)
   d. (−1,−2)
   e. (2,−1)

26. Simplify $(2x^4)(-3x)^2$.

27. Multiply the two monomials $(-7xy^3)(8x^2y^4)$.

28. Factor $24a^2b^3 - 32ab^4$.

29. Factor $x^2 - x - 12$.

30. If $9a + 5 = -22$, what is the value of a?

31. Multiply $(-6p^3q)^2(-2pq^5)$.

32. Multiply the binomials $(2x + 3)(x - 6)$.

33. Multiply the binomials $(3a - 4b)(4a + 3b)$.

34. Factor $18a^2b^2c^2 + 54a^3b^2c^3$.

35. A function is a special kind of relation in which
   a. no range value is repeated because the function would fail the vertical line test.
   b. each member of the range corresponds with exactly one member of the domain.
   c. each member of the domain corresponds with exactly one member of the range.
   d. each member of the domain corresponds with at least two members of the range.

36. Which of the following statements accurately reflects the inequality $2x - 4 < 7(x - 2)$?

a. Seven and the quantity two less than a number is greater than four less than two times the number.

b. The product of seven and the quantity two less than a number is greater than four less than two times the number.

c. The product of seven and the quantity two less than a number is less than four less than two times the number.

d. The product of seven and the quantity two less than a number is greater than four less than two more than the number.

37. Factor the trinomial $x^2 - 10 + 24$.

38. For the equation $f(x) = |x| + 3$, what is the range?

39. Each term in the following sequence is -4 times the previous term. What is the value of xy?

$x, y, -64, 256, \ldots$

40. Each term in the following sequence is three times the previous term. What is the product of the 100th and 101st terms of the sequence?

$1, 3, 9, 27, \ldots$

41. Each term in the following sequence is two less than three times the previous term. What is the next term of the sequence?

$-1, -5, -17, -53, \ldots$

42. Each term in the following sequence is equal to the sum of the two previous terms.

$\ldots a, b, c, d, e, f, \ldots$

All of the following are equal to the value of d EXCEPT

a. $e - c$

b. $b + c$

c. $a + 2b$

d. $e - 2b$

e. $f - e$

43. Given the following system of equations, what is one possible value of m?

$m(n + 1) = 2$

$m - n = 0$

44. Given the following system of equations, what is the value of $\frac{c}{d}$?

$\frac{c - d}{5} - 2 = 0$

$c - 6d = 0$

45. Given the following system of equations, what is the value of $a + b$?

$4a + 6b = 24$

$6a - 12b = -6$

46. Given the following system of equations, what is the value of $x - y$?

$\frac{x}{3} - 2y = 14$

$2x + 6y = -6$

47. Find the value of the inequality $3x - 6 \leq 4(x + 2)$ in terms of x.

48. If $6x - 4x + 9 = 6x + 4 - 9$, what is the value of x?

49. What is the solution set for the inequality $-8(x + 3) \leq 2(-2x + 10)$?

50. If $\frac{3c^2}{6c} + 9 = 15$, what is the value of c?

ANSWERS

1. First compare the given equation to the $y = ax^2 + bx + c$ formula:

   $y = ax^2 + bx + c$

   $y = 2x^2 - 8x + 1$

   The a and the c are clear, but to clearly see what b equals, convert the subtraction to add the opposite:

   $y = 2x^2 + (-8)x + 1$

   Thus, $a = 2$, $b = -8$, and $c = 1$.

   The x-coordinate of the turning point, or vertex, of the parabola is given by:

   $x = \frac{-b}{2a}$

   Substitute in the values from the equation:

   $x = \frac{-b}{2a} = \frac{-(-8)}{(2)(2)} = \frac{8}{4} = 2$

   When $x = 2$, y will be:

   $y = (2)(2)^2 - (8)(2) + 1$

   $y = (2)(4) - (2)(8) + 1$

   $= 8 - 16 + 1$

   $= -8 + 1$

   $= -7$

   Thus, the coordinates of the vertex equal $(2,-7)$. For more help with this concept, see Chapters 7 and 8.

2. First compare the given equation to the $y = ax^2 + bx$ formula:

   $y = ax^2 + bx + c$

   $y = 2x^2 + 16x + 0$

   Thus, $a = 2$, $b = 16$, and $c = 0$.

   The x-coordinate of the vertex is given by:

   $x = \frac{-b}{2a}$

   Substitute in the values from the equation:

   $x = \frac{-b}{2a} = \frac{-16}{4} = -4$

   Thus, the line of symmetry is $x = -4$.

   When $x = -4$, y will be:

   $y = (2)(-4)^2 + (16)(-4)$

   $y = (2)(16) + (16)(-4)$

   $= 32 - 64$

   $= -32$

The vertex is at $(-4,-32)$. For more help with this concept, see Chapters 7 and 8.

3. First compare the given equation to the $y = ax^2 + bx + c$ formula:

$y = ax^2 + bx + c$

$y = 5x^2 + 0(x) + 0$

Thus, $a = 5$, $b = 0$, and $c = 0$.

The x-coordinate of the turning point, or vertex, of the parabola is given by:

$x = \frac{-b}{2a}$

Substitute in the preceding values:

$x = \frac{-b}{2a} = \frac{-(0)}{2(5)} = 0$

When $x = 0$, y will be:

$y = (5)(0^2)$

$y = 0$

The vertex is $(0,0)$. For more help with this concept, see Chapters 7 and 8.

4. Because the rule of the sequence is that each term is nine more than $\frac{1}{3}$ the previous term, to find the value of x, multiply the last term, 36, by $\frac{1}{3}$, then add 9: $(36)(\frac{1}{3}) = 12$, $12 + 9 = 21$. In the same way, the value of y is $21(\frac{1}{3}) + 9 = 7 + 9 = 16$. Therefore, the value of $y - x = 16 - 21 = -5$. For more help with this concept, see Chapter 9.

5. Because the rule of the sequence is each term is 20 less than five times the previous term, to find the value of x, add 20 to 0 and divide by 5: $\frac{(0 + 20)}{5} = \frac{20}{5} = 4$. In the same way, the value of y is $\frac{(-120 + 20)}{5} = \frac{-100}{5} = -20$. Therefore, the value of $x + y = 4 + -20 = -16$. For more help with this concept, see Chapter 9.

6. Continue the sequence: 28.5 is the fourth term of the sequence. The fifth term is $\frac{28.5}{2} - 2 = 14.25 - 2 = 12.25$. The sixth term is $\frac{12.25}{2} - 2 = 6.125 - 2 = 4.125$, the seventh term is $\frac{4.125}{2} - 2 = 2.0625 - 2 = 0.0625$. Half of this number minus two will yield a negative value, so the eighth term of the sequence is the first term of the sequence that is a negative number. For more help with this concept, see Chapter 9.

7. Because the rule of the sequence is each term is 16 more than -4 times the previous term, to find the value of y, subtract 16 from -80 and divide by -4: $\frac{(-80 - 16)}{-4} = \frac{-96}{-4} = 24$. In the same way, the value of x is $\frac{24 - 16}{-4} = \frac{8}{-4} = -2$. Therefore, the value of $x + y = -2 + 24 = 22$. For more help with this concept, see Chapter 9.

8. To isolate the variable, subtract 15 from both sides:

$5\sqrt{c} + 15 - 15 = 35 - 15$

$5\sqrt{c} = 20$

Next, divide both sides by 5:

$\frac{5\sqrt{c}}{5} = \frac{20}{5}$

$\sqrt{c} = 4$

Last, square both sides:

$(\sqrt{c})^2 = 4^2$

$c = 16$

For more help with this concept, see Chapters 3 and 7.

9. $A = \begin{bmatrix} 3 & -1 & 2 \\ 0 & 2 & -4 \end{bmatrix}$

So, $-A = \begin{bmatrix} -3 & 1 & -2 \\ 0 & -2 & 4 \end{bmatrix}$

For more help with this concept, see Chapter 6.

10. Factor each term in the numerator: $\sqrt{(a^2b)} = (\sqrt{(a^2)})(\sqrt{b}) = a\sqrt{b}$; $\sqrt{(ab^2)} = (\sqrt{a})(\sqrt{(b^2)}) = b\sqrt{a}$. Next, multiply the two radicals. Multiply the coefficients of each radical and multiply the radicands of each radical: $(a\sqrt{b})(b\sqrt{a}) = ab\sqrt{ab}$. The expression is now $ab\frac{\sqrt{ab}}{\sqrt{ab}}$. Cancel the $\sqrt{ab}$ terms from the numerator and denominator,

leaving ab. For more help with this concept, see Chapters 3 and 7.

11. First, cube the ab term: $(ab)^3 = a^3b^3$. Next, raise the fraction $\frac{(a^3b^3)}{b}$ to the fourth power. Multiply each exponent of the a and b terms by 4: $(\frac{(a^3b^3)}{b})^4 = \frac{(a^{12}b^{12})}{b^4}$. To divide b^{12} by b^4, subtract the exponents: $\frac{b^{12}}{b^4} = b^8$. Therefore, $\frac{(a^{12}b^{12})}{b^4} = a^{12}b^8$. For more help with this concept, see Chapter 7.

12. First, cube the $4g^2$ term. Cube the constant, 4, and multiply the exponent of g, 2, by 3: $(4g^2)^3 = 64g^6$. Next, multiply $64g^6$ by g^4. Add the exponents of the g terms: $(64g^6)(g^4) = 64g^{10}$. Finally, take the square root of $64g^{10}$: $(64g^{10})^{\frac{1}{2}} = 8g^5$, because $(8g^5)(8g^5) = 64g^{10}$. For more help with this concept, see Chapter 7.

13. Because $(a^{\frac{2}{3}})^2 = a^{\frac{4}{3}}$, the value of $a^{\frac{4}{3}}$ is equal to the value of $a^{\frac{2}{3}}$ squared. Therefore, $a^{\frac{4}{3}} = 6^2 = 36$. For more help with this concept, see Chapter 7.

14. $(\sqrt{p})^4 = (p^{\frac{1}{2}})^4$. Multiply the exponents: $(p^{\frac{1}{2}})^4 = p^2$. Substitute $-\frac{1}{3}$ for q:

    $p^2 = (-\frac{1}{3})^{-2}$. A fraction with a negative exponent can be rewritten as the

    reciprocal of the fraction with a positive exponent: $(-\frac{1}{3})^{-2} = (-3)^2 = 9$;

    $p^2 = 9$, and $p = -3$ or 3. For more help with this concept, see Chapter 7.

15. Substitute $\frac{1}{3}$ for a and 9 for b: $\left(\frac{1}{3\sqrt{9}}\right) = (\frac{1}{3})(3) = 1$; 1 is raised to the

    power of 3, but the value of the exponent does not matter; 1 raised to

    any power is 1. For more help with this concept, see Chapter 7.

16. Substitute 20 for n: $\frac{\sqrt{20 + 5}}{\sqrt{20}}\left(\frac{20}{2}\sqrt{5}\right) = \frac{\sqrt{25}}{\sqrt{20}}(10\sqrt{5}) = \frac{5}{2\sqrt{5}}(10\sqrt{5})$.

    Cancel the $\sqrt{5}$ terms and multiply the fraction by 10: $\frac{5}{2\sqrt{5}}(10\sqrt{5}) =$

    $\frac{5(10)}{2} = \frac{50}{2} = 25$. For more help with this concept, see Chapter 7.

17. If $a^2 = b = 4$, then $a = 2$. Substitute 2 for a and 4 for b: $\left[\frac{4\sqrt{4}}{(2)^4}\right]^2 = \left[\frac{(4)(2)}{16}\right]^2$

    $= (\frac{8}{16})^2 = (\frac{1}{2})^2 = \frac{1}{4}$. For more help with this concept, see Chapter 7.

18. For $x - y > 5$, rearrange so that the y is by itself on the left: $x - y > 5$ becomes $x > 5 + y$, which equals $x - 5 > y$, or $y < x - 5$. For $y < x - 5$, the slope is 1 and the y-intercept is -5. Start at the y-intercept $(0,-5)$ and go up 1 and over 1 (right) to plot points. This line will be dashed because the symbol is $<$. The area below this line satisfies the condition.

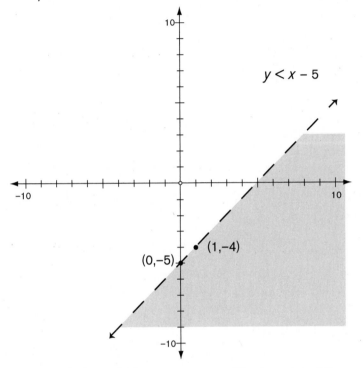

For more help with this concept, see Chapters 4 and 5.

19. For $3x - y < 2$, rearrange so that the y is by itself on the left: $3x - y < 2$ becomes $-y < -3x + 2$, which equals $y > 3x - 2$. For $y > 3x - 2$, the slope is 3 and the y-intercept is -2. Start at the y-intercept $(0,-2)$ and go up 3 and over 1 (right) to plot points. This line will be dashed because the symbol is $>$. The area above this line satisfies the condition.

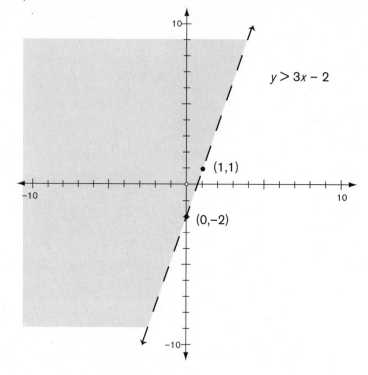

For more help with this concept, see Chapter 4.

20. For $y = 3x - 2$, the slope is 3 and the y-intercept is -2. Start at the y-intercept $(0,-2)$ and go up 3 and over 1 (right) to plot points.

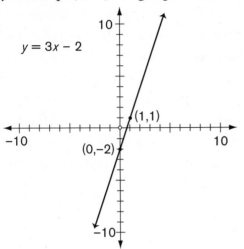

For more help with this concept, see Chapters 4 and 5.

21. Rearrange the first equation so that the y is by itself on the left: $6x + 3y < 9$ becomes $3y < -6x + 9$, which equals $y < -2x + 3$. For $y < -2x + 3$, the slope is -2 and the y-intercept is 3. Start at the y-intercept $(0,3)$ and go down 2 and over 1 (right) to plot points. This line will be dashed because the symbol is $<$. The area below this line satisfies the condition. Rearrange the second equation so that the y is by itself on the left: $-x + y \leq -4$ becomes $y \leq x + -4$. For $y \leq x + -4$, the slope is 1 and the y-intercept is 4. Start at the y-intercept $(0,4)$ and go up 1 and over 1 (right) to plot points. This line will be solid because the symbol is $\leq$. The area below this line satisfies the condition. Shade the area that satisfies both conditions:

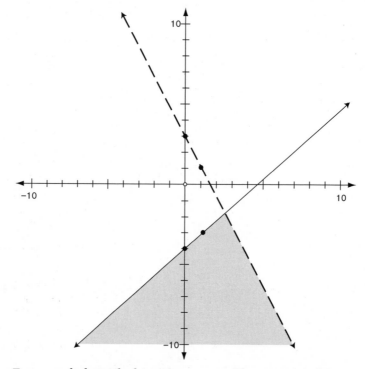

For more help with this concept, see Chapters 4 and 5.

22. For $y \geq 6$, draw a solid line at $y = 6$. The area above this line satisfies the given condition. For $x + y < 3$, rearrange so that the y is by itself on the left: $x + y < 3$ becomes $y < -x + 3$. For $y < -x + 3$, the slope is -1 and the y-intercept is 3. Start at the y-intercept $(0,3)$ and go down 1 and over 1 (right) to plot points. This line will be dashed because the symbol is $<$. The area below this line satisfies the condition. Shade the area that satisfies both conditions:

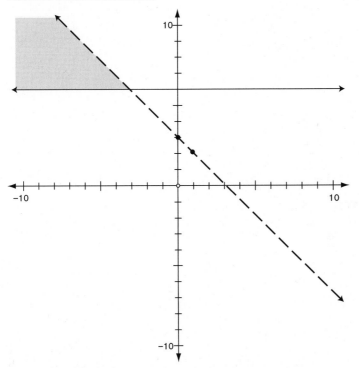

For more help with this concept, see Chapters 4 and 5.

23. For $x \leq -2$, draw a solid line at $x = 2$. The area to the left satisfies the condition, so shade it:

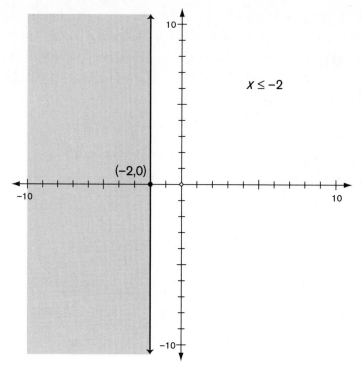

24. For $y \geq 0$, draw a solid line at $y = 0$. The area above the line satisfies the condition, so shade it:

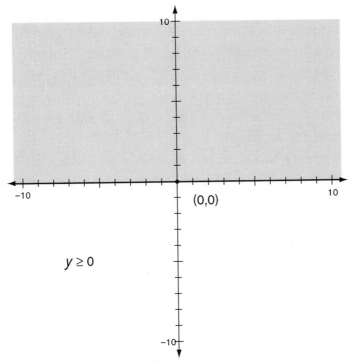

For more help with this concept, see Chapters 4 and 5.

25. Plugging in $(1,1)$ gives $9 > 9$, which is not a true statement.
Plugging in $(2,1)$ gives $9 > 15$, which is not a true statement.
Plugging in $(1,2)$ gives $6 > 9$, which is not a true statement.
Plugging in $(-1,-2)$ gives $18 > -3$, which IS a true statement.
Plugging in $(2,-1)$ gives $15 > 15$, which is not a true statement.
Therefore, $(-1,-2)$ is the only correct solution.
For more help with this concept, see Chapters 4 and 5.

26. $18x^6$; $-3x$ is squared properly to $9x^2$ and the exponents 4 and 2 are properly added to 6. For more help with this concept, see Chapters 3 and 7.

27. $-56x^3y^7$; -56 is the proper product of -7 and 8 and the exponents are properly added. For more help with this concept, see Chapters 3 and 7.

28. $8ab^3(3a - 4b)$; this answer has the correct greatest common factor (GCF) and is properly factored. For more help with this concept, see Chapter 7.

29. $(x + 3)(x - 4)$; after FOIL, all three terms would be correct. For more help with this concept, see Chapter 7.

30. To solve the equation, subtract 5 from both sides of the equation, then divide by 9: $9a + 5 = -22$, $9a + 5 - 5 = -22 - 5$, $9a = -27$, $a = -3$. For more help with this concept, see Chapter 3.

31. $-72p^7q^7$; the product of $(-6)^2$ and -2 is -72, and there are seven factors of both p and q. For more help with this concept, see Chapter 3.

32. $2x^2 - 9x - 18$; this is the result of the FOIL method. For more help with this concept, see Chapter 7.

33. $12a^2 - 7ab - 12b^2$; this is the result of the FOIL method. For more help with this concept, see Chapter 7.

34. $18a^2b^2c^2 (1 + 3ac)$; this is the proper factorization using the GCF. By redistributing, the original binomial can be obtained. For more help with this concept, see Chapter 7.

35. Choice **c** is the definition of a function. Choice **a** is incorrect; if a range value is repeated, the function would not fail the vertical line test. Choice **b** has the words *domain* and *range* reversed. Choice **d** is incorrect; this guarantees that the relation would *not* be a function. For more help with this concept, see Chapter 4.

36. Choice **b** is correct; it indicates the correct translation of the inequality. Choice **a** is the equation $2x - 4 < 7 + (x - 2)$. Choice **c** represents $2x - 4 > 7(x - 2)$, and choice **d** is the equation $2 + x - 4 < 7(x - 2)$. For more help with this concept, see Chapter 3.

37. $(x - 4)(x - 6)$; This is one possible factoring for the given trinomial. For more help with this concept, see Chapter 8.

38. All real numbers greater than or equal to 3; the smallest value possible for $|x|$ is 0, which means that the smallest possible range is equal to 3. For more help with this concept, see Chapter 4.

39. Because each term in the sequence is -4 times the previous term, y is equal to $\frac{-64}{-4} = 16$, and $x = \frac{16}{-4} = -4$. Therefore, $xy = (16)(-4) = -64$. For more help with this concept, see Chapter 9.

40. Every term in the sequence is 3 raised to a power. The first term, 1, is 3^0. The second term, 3, is 3^1. The value of the exponent is one less than the position of the term in the sequence. The 100th term of the sequence is equal to $3^{100-1} = 3^{99}$ and the 101st term in the sequence is equal to $3^{99+1} = 3^{100}$. To multiply two terms with common bases, add the exponents of the terms: $(3^{99})(3^{100}) = 3^{199}$. For more help with this concept, see Chapter 9.

41. Because the rule of the sequence is each term is two less than three times the previous term, multiply the last term, -53, by 3, then subtract 2: $(-53)(3) = -159 - 2 = -161$. For more help with this concept, see Chapter 9.

42. **d.**; because each term in the sequence is equal to the sum of the two previous terms, $d = b + c$. You know that $e = c + d$, because c and d are the two terms previous to e. If $e = c + d$, then, by subtracting c from both sides of the equation, $d = e - c$. In the same way, $f = d + e$, the terms that precede it, and that equation can be rewritten as $d = f - e$; $d = b + c$, and $c = a + b$. Therefore, $d = b + (a + b)$, $d = a + 2b$. However, d is not equal to $e - 2b$. Also, $d = e - c$, and $c = a + b$, not $2b$, because a is not equal to b. For more help with this concept, see Chapter 9.

43. Solve the second equation for n in terms of m: $m - n = 0$, $n = m$. Substitute this expression for n in the first equation and solve for m:

$m(n + 1) = 2$

$m(m + 1) = 2$

$m^2 + m = 2$

$m^2 + m - 2 = 0$

$(m + 2)(m - 1) = 0$

$m + 2 = 0$, $m = -2$

$m - 1 = 0$, $m = 1$

For more help with this concept, see Chapter 5.

44. Solve the second equation for c in terms of d: $c - 6d = 0$, $c = 6d$. Substitute this expression for c in the first equation and solve for d:

$\frac{c - d}{5 - 2} = 0$

$\frac{6d - 2}{5 - 2} = 0$

$\frac{5d}{5} - 2 = 0$

$d - 2 = 0$

$d = 2$

Substitute the value of d into the second equation and solve for c:

$c - 6(2) = 0$

$c - 12 = 0$

$c = 12$

Because $c = 12$ and $d = 2$, the value of $\frac{c}{d} = \frac{12}{2} = 6$.

For more help with this concept, see Chapter 5.

45. Divide the second equation by 2 and add it to the first equation. The b terms will drop out, and you can solve for a:

$$\frac{6a - 12b}{2} = \frac{-6}{2} = 3a - 6b = -3$$

$$4a + 6b = 24$$
$$+\ 3a - 6b = -3$$
$$\overline{7a = 21}$$
$$a = 3$$

Substitute the value of a into the first equation and solve for b:

$$4(3) + 6b = 24$$
$$12 + 6b = 24$$
$$6b = 12$$
$$b = 2$$

Because $a = 3$ and $b = 2$, the value of $a + b = 3 + 2 = 5$.

For more help with this concept, see Chapter 5.

46. Multiply the first equation by -6 and add it to the second equation. The x terms will drop out, and you can solve for y:

$$-6(\tfrac{x}{3} - 2y = 14) = -2x + 12y = -84$$

$$-2x + 12y = -84$$
$$+\ 2x +\ 6y = -6$$
$$\overline{18y = -90}$$
$$y = -5$$

Substitute the value of y into the second equation and solve for x:

$$2x + 6(-5) = -6$$
$$2x - 30 = -6$$
$$2x = 24$$
$$x = 12$$

Because $x = 12$ and $y = -5$, the value of $x - y = 12 - (-5) = 12 + 5 = 17$.

For more help with this concept, see Chapter 5.

47. First, multiply $(x + 2)$ by 4: $4(x + 2) = 4x + 8$. Then, subtract $3x$ from both sides of the inequality and subtract 8 from both sides of the inequality:

$$3x - 6 \leq 4x + 8$$
$$3x - 6 - 3x \leq 4x + 8 - 3x$$
$$-6 \leq x + 8$$
$$-6 - 8 \leq x + 8 - 8$$
$$x \geq -14$$

For more help with this concept, see Chapter 3.

48. First, combine like terms on each side of the equation: $6x - 4x = 2x$ and $4 - 9 = -5$. Now, subtract $2x$ from both sides of the equation and add 5 to both sides of the equation:

$2x + 9 = 6x - 5$

$2x - 2x + 9 = 6x - 2x - 5$

$9 = 4x - 5$

$9 + 5 = 4x - 5 + 5$

$14 = 4x$

Finally, divide both sides of the equation by 4: $\frac{14}{4} = \frac{4x}{4}, x = \frac{14}{4} = \frac{7}{2}$. For more help with this concept, see Chapter 3.

49. First, multiply $(x + 3)$ by -8 and multiply $(-2x + 10)$ by 2: $-8(x + 3) = -8x - 24, 2(-2x + 10) = -4x + 20$. Then, add $8x$ to both sides of the inequality and subtract 20 from both sides of the inequality:

$-8x - 24 \leq -4x + 20$

$-8x - 24 + 8x \leq -4x + 20 + 8x$

$-24 \leq 4x + 20$

$-24 - 20 \leq 4x + 20 - 20$

$-44 \leq 4x$

Finally, divide both sides of the inequality by 4: $\frac{-44}{4} \leq \frac{4x}{4}, x \geq -11$. For more help with this concept, see Chapter 3.

50. First, reduce the fraction $\frac{3c^2}{6c}$ by dividing the numerator and denominator by $3c$: $\frac{3c^2}{6c} = \frac{c}{2}$. Now, subtract 9 from both sides of the equation and then multiply both sides of the equation by 2:

$\frac{c}{2} + 9 = 15$

$\frac{c}{2} + 9 - 9 = 15 - 9$

$\frac{c}{2} = 6$

$(2)(\frac{c}{2}) = (6)(2)$

$c = 12$

For more help with this concept, see Chapter 3.

Glossary

Absolute value: The distance a number or expression is from zero on a number line.

Additive property of zero: When you add zero to a number, the result is that number. For example, $-6 + 0 = 6$ or $x + 0 = x$.

Associative property of addition: When adding three or more addends, the grouping of the addends does not affect the sum.

Associative property of multiplication: When multiplying three or more factors, the grouping of the factors does not affect the product.

Average: The quantity found by adding all the numbers in a set and dividing the sum by the number of addends; also known as the *mean*.

Base: A number used as a repeated factor in an exponential expression. In 5^7, 5 is the base.

Binomial: A polynomial with two unlike terms, such as $2x + 4y$.

Circumference: The distance around a circle.

Coefficient: A number placed next to a variable.

Common factors: The factors shared by two or more numbers.

Common multiples: Multiples shared by two or more numbers.

Commutative property: Allows you to change the order of the numbers when you add or multiply.

Commutative property of addition: When using addition, the order of the addends does not affect the sum.

Commutative property of multiplication: When using multiplication, the order of the factors does not affect the product.

Composite number: A number that has more than two factors.

Coordinate plane: A grid divided into four quadrants by both a horizontal x-axis and a vertical y-axis.

Coordinate points: Points located on a coordinate plane.

Cross product: A product of the numerator of one fraction and the denominator of a second fraction.

Denominator: The bottom number in a fraction. For example, 7 is the denominator of $\frac{3}{7}$.

Difference: The result of subtracting one number from another.

Distributive property: When multiplying a sum (or a difference) by a third number, you can multiply each of the first two numbers by the third number and then add (or subtract) the products.

Dividend: A number that is divided by another number. In the division problem $a\overline{)b}$, the quantity b is the dividend.

Divisor: A number that is divided into another number. In the division problem $a\overline{)b}$, the quantity a is the divisor.

Domain: All the x-values of a function.

Equation: A mathematical statement that contains an equal sign.

Evaluate: Substitute a number for each variable and simplify.

Even number: A whole number that can be divided evenly by the number 2 resulting in a whole number.

Exponent: A number that tells you how many times a number, the base, is a factor in the product. In 5^7, 7 is the exponent.

Factor: A number that is multiplied to find a product.

Function: A relationship in which one value depends upon another value.

Geometric sequence: A sequence that has a constant ratio between terms.

Greatest common factor: The largest of all the common factors of two or more numbers.

Improper fraction: A fraction in which the numerator is greater than or equal to the denominator. A fraction greater than or equal to 1.

Inequality: Two expressions that are not equal and are connected with an inequality symbol such as $<$, $>$, $\leq$, $\geq$, or $\neq$.

Integers: Positive or negative whole numbers and the number zero.

Irrational numbers: Numbers that cannot be expressed as terminating or repeating decimals.

Least common denominator (LCD): The smallest number divisible by two or more denominators.

Least common multiple (LCM): The smallest of all the common multiples of two or more numbers.

Like terms: Two or more terms that contain the exact same variables.

Line: A straight path that continues infinitely in two directions. The geometric notation for a line through points A and B is $\overleftrightarrow{AB}$.

Linear equation: An equation with two variables that describes a line. The variable in a linear equation cannot contain an exponent greater than one. It cannot have a variable in the denominator, and the variables cannot be multiplied.

Matrix: A rectangular array of numbers.

Mean: The quantity found by adding all the numbers in a set and dividing the sum by the number of addends; also known as the *average*.

Median: The middle number in a set of numbers arranged from least to greatest.

Mode: The number that occurs most frequently in a set of numbers.

Monomial: A polynomial with one term, such as $5b^6$.

Multiple: A number that can be obtained by multiplying a number x by a whole number.

Negative number: A number less than zero.

Numerator: The top number in a fraction. For example, 3 is the numerator of $\frac{3}{7}$.

Odd number: A number that cannot be divided evenly by the number 2.

Order of operations: The order of performing operations to get the correct answer. The order you follow is:
1. Simplify all operations within grouping symbols such as parentheses, brackets, and braces.
2. Evaluate all exponents.
3. Do all multiplication and division in order from left to right.
4. Do all addition and subtraction in order from left to right.

Ordered pair: A location of a point on the coordinate plane in the form of (x,y). The x represents the location of the point on the horizontal x-axis, and the y represents the location of the point on the vertical y-axis.

Origin: coordinate point $(0,0)$; the point on a coordinate plane at which the x-axis and y-axis intersect.

Parallel lines: Two lines in a plane that do not intersect.

Percent: A ratio that compares a number to 100; 45% is equal to $\frac{45}{100}$.

Perfect square: A whole number whose square root is also a whole number.

Perimeter: The distance around a figure.

Permutation: The arrangement of a group of items in a specific order.

Perpendicular lines: Lines that intersect to form right angles.

Polynomial: The sum of distinct terms, each of which consists of a number, one or more variables raised to an exponent, or both.

Positive number: A number greater than zero.

Prime factorization: The process of breaking down factors into prime numbers.

Prime number: A number that has only 1 and itself as factors.

Product: The result of multiplying two or more factors.

Proper fraction: A fraction whose numerator is less than its denominator; a fraction less than 1.

Proportion: An equality of two ratios in the form $\frac{a}{b} = \frac{c}{d}$.

Quadrants: The coordinate plane is divided into four equal parts called quadrants. A number names each quadrant. The quadrant in the upper-right-hand quadrant is Quadrant I. You proceed counterclockwise to name the other quadrants.

Quadratic equation: An equation in the form $ax^2 + bx + c = 0$, where a, b, and c are numbers and $a \neq 0$.

Quadratic trinomial: An expression that contains an x^2 term, x term, and a constant.

Quotient: The result of dividing two or more numbers.

Radical equation: An equation that has a variable in the radicand.

Radical sign: The mathematical symbol that tells you to take the root of a number.

Radicand: The number under the radical sign in a radical.

Radius: A line segment inside a circle with one point on the radius and the other point at the center on the circle. The radius is half the diameter. This term can also be used to refer to the length of such a line segment. The plural of *radius* is *radii*.

Range: All the solutions to $f(x)$ in a function.

Ratio: A comparison of two quantities measured in the same units.

Rational numbers: All numbers that can be written as fractions, terminating decimals, and repeating decimals.

Reciprocals: Two numbers whose product is 1. For example, $\frac{5}{4}$ is the reciprocal of $\frac{4}{5}$.

Set: A collection of certain numbers.

Simplify: To combine like terms and reduce an equation to its most basic form.

Slope: The steepness of a line, as determined by vertical change/horizontal change, or $\frac{y_2 - y_1}{x_2 - x_1}$, on a coordinate plane where (x_1, y_1) and (x_2, y_2) are two points on that line.

Slope-intercept form: $y = mx + b$. Also known as $y =$ form.

Square of a number: The product of a number and itself, such as 6^2, which is 6×6.

Square root: One of two equal factors whose product is the square, such as $\sqrt{7}$.

Sum: The result of adding one number to another.

System of equations: Two or more equations with the same variables.

System of inequalities: Two or more inequalities with the same variables.

Trinomial: A polynomial with three unlike terms, such as $y^3 + 8z - 2$.

Variable: A letter that represents an unknown number.

Vertex: A point at which two lines, rays, or line segments connect.

Whole numbers: The counting numbers: 0, 1, 2, 3, 4, 5, 6, . . .

x-axis: The horizontal line that passes through the origin on the coordinate plane.

y-axis: The vertical line that passes through the origin on the coordinate plane.

y-intercept: Point where the line intersects the-axis.

Zero product rule: If the product of two or more factors is 0, then at least one of the factors is 0.

Notes

Notes